Women Transforming the Landscape of Science and Tech

Challenges, Triumphs, and Vision for the Future

Women Transforming the Landscape of Science and Tech

Challenges, Triumphs, and Vision for the Future

curated by

Cathy Derksen

Action Takers Publishing Inc

San Diego, California

©Copyright 2023 by Action Takers Publishing Inc™

All rights reserved. No part of this publication may be reproduced or transmitted in any form or by any means, mechanical or electronic, including photocopying and recording, or by any information storage and retrieval system, without permission in writing from the author or publisher (except by reviewer, who may quote brief sections and/or show brief video clips in a review).

Disclaimer: The Publisher and the Author make no representations or warranties with respect to the accuracy or completeness of the contents of this work and specifically disclaim all warranties, including without limitation warranties of fitness for a particular purpose. No warranty may be created or suitable for every situation. This works is sold with the understanding that the Publisher is not engaged in rendering legal, accounting, or other professional services. If professional assistance is required, the services of a competent professional person should be sought.

Neither the Publisher nor the Author shall be liable for damages arising herefrom. The fact that an organization or website is referred to in this work as a referred source of further information does not mean that the Author or the Publisher endorses the information the organization or website may provide or recommendations it may make. Further, readers should be aware that websites listed in this work may have changed or disappeared between when this work was written and when it was read.

ISBN # (paperback) 978-1-956665-26-0

ISBN # (Kindle) 978-1-956665-27-7

Published by Action Takers Publishing Inc

Email: support@actiontakerspublishing.com

Website: www.actiontakerspublishing.com

100% of the net proceeds of the sales of this book will be donated to a 501(c)(3) nonprofit charity

Cover Design by Sam Art Studio

Printed in the United States of America

Table of Contents

Introduction

The vast array of fields in science and technology hold an amazing collection of fascinating areas of study and career paths. Women and men alike are drawn to these disciplines, but far too often the journey for women can include extra obstacles. This book is filled with a collection of stories shared by women who have faced these situations. You will hear about their triumphs and challenges, as well as their vision for the future. These authors represent a global community of women in a broad range of careers in STEM (Science Tech Engineering and Math).

My intention with this book is to give a platform to lift the voices of these women. To motivate the women currently in STEM fields to take on leadership roles and support discussions around diversity, inclusion, and equity. To give women the courage to follow their dreams and reach for the big vision they have for themselves and others. To inform and support younger women to see a place for themselves in this enormous area of opportunities.

The members of this amazing international team of authors have all embraced courage in their own way and have felt compelled to share their journey here with you.

I would like to thank each author in this book for taking on this challenge. Sharing your stories with the world takes enormous courage. Sharing your voices as women in STEM can lift the whole community.

I am honored to have been entrusted with their work and I am proud to be part of each one's individual journey into growing as an author.

Cathy Derksen

CHAPTER 1

Science Is for Everyone

Carrie Boyce

Science is for everyone. At least, that's the dream. I've seen it printed on more inspiring posters and tote bags than I care to count. But the reality is that for many, science remains an unwelcoming space, entrenched in the echoes of cis, white, heteronormative, and patriarchal structures.

Since this is a book about women's experiences in STEM, I'll start with a hypothesis: I suspect every author in this book has at least one story describing how they've been made to feel excluded from science. I bet you'll even read about some of them.

The fact that this is hardly news is damning in and of itself. How, in the 21st century, with all of the equity, diversity, and inclusivity initiatives I keep hearing about, are we *still* excluding billions of people around the world from science because of their disabilities or how they look, dress, speak, or love? How are we still holding back the breadth and impact of scientific discovery?

I've decided the behemoth of academia is out of date. It's like an outdated piece of software from the '80s. Sure, the infrastructure has evolved around it, but rather than rebuilding the operating system for this new global context, we just keep adding patches, hoping they'll address the vulnerabilities. They won't. They can't possibly keep up with the rate of change. And so we find ourselves stuck, facing error message after error message and wondering why our updates to the code aren't solving the problem.

This lack of representation in who does research in turn impacts the quality of scientific research. For example, too many studies conflate gender and sex, so they get communicated inaccurately. Binary collection of study participant data leads to less accuracy in interpreting the outputs – how can we purport to be developing precision medicine if all of our genomics data is annotated in the binary? There's a lack of representation in clinical trials for those on hormone therapies (e.g. those transitioning or going through menopause) and advocacy for important issues like reproductive rights is often still exclusionary, which ultimately leads to more harm to trans and queer people.

To truly have gender-inclusive and representative research, we need to ensure our research and related advocacy is inclusive to all genders.

Since I'm queer, I'm going to share a few things you might not know if you're straight in science.

I'll start generally. A paper in Science Advances recently revealed that, "Over half of Americans still harbor some level of prejudice toward non-heterosexual, transgender, and gender nonbinary persons, and those prejudices often translate into overtly or subtly biased treatment of LGBTQ colleagues." Let's be real: it's never too long between news cycles before another horror story of violence and discrimination against members of the queer community presents itself.

But across the board, members of the 2SLGBTQIA+ community working in science are reporting being on the receiving end of exclusionary, offensive, or harassing behaviour at work. A UK study of physical scientists revealed that nearly 30% of queer scientists and half of transgender scientists had considered leaving the workplace because of unfriendly or hostile climates and discrimination. In Canada, where I currently live, the research barely exists.

Despite the expansion of queer rights over the past twenty years and despite the presumed objectivity and universality of STEM, queer-identifying people continue to face a multitude of biases in STEM education, in STEM academia, and in STEM-related workplaces. Science remains very straight.

So when then Ph.D. candidates Dr. Samantha Yammine (better known as Science Sam on social media), Dr. Shawn Hercules and Dr. Geith Maal-Bared were gathered round a table in a bar watching RuPaul's Drag Race and lamenting how science had made them all feel "too much" – too loud, too bubbly, and too feminine are all adjectives they'd been labeled with in science – they knew they wanted to rebel. The answer? A science-themed drag show.

RuPaul probably described it best when they said, "we're all born naked and the rest is drag!" Everything about how we present ourselves – no matter how you identify – is a form of drag and as an artform exaggerates and bends gender expression. Today, drag shows remain a beloved touchstone of 2SLGBTQIA+ communities.

When Samantha brought the idea to me in my then role as Programs Manager at the Royal Canadian Institute for Science (RCIScience), I knew the answer was yes before she even finished the pitch. I don't think an idea has ever appealed to me more. Here was a chance to call out science – and invite it to call us in.

We immediately got to work and in 2019 hosted the world's first fully science-themed drag show – where we take scientists, dress them in Drag (often for the first time), and train them to share research with the public in queer-friendly spaces.

Our kings and queens serve drama and science in equal measure, turning live DNA extractions into burlesque performances, highlighting the climate crisis with lip syncs and fiery costume reveals, challenging people's perceptions of bees using Pictionary and showing us the correct techniques for flossing using feather boas. They're giving us stand up, they're giving us dance demos, they're sharing their research in ways we actually remember because this show is anything but painting by numbers (probably because we spend so much time painting our faces).

More than entertainment, Science is a Drag takes an intersectional approach to challenge the cis-heteronormative structures and stereotypes in STEM, celebrating our differences as sources of enhanced resourcefulness. It provides an empowering and inclusive platform for queer scientists and communicators to share their passions for STEM in an unapologetically queer manner. And it fosters a safe and accessible space for the queer community and its allies to engage with science without fear of exclusion or judgment.

The pandemic may have stalled our expansion, but in a handful of shows, we've already reached hundreds of people in person, tens of thousands more online, and been featured in local and international publications.

I've spent most of my career working in science engagement, trying to get the public to feel comfortable actively interacting with

STEM. The tricky thing is reaching beyond the converted. Of all the public talks, science festivals and demonstrations I've hosted in the street, it's tough to really believe that you're reaching the science averse. Imagine my surprise and delight when over half of our audience reported that they'd never attended a science-themed event in the past and would not have come to our show were it not a drag event. Almost 90% feel more comfortable at Science is a Drag than typical science events and, unsurprisingly, they *all* want to come back for more. More than that, they are. Because it's the emotional impact of engaging with science at our shows that hits different, as audience members and performers alike connect parts of their identities together and embrace who they are fully. But don't just take my word for it. In their own words:

"Attending Science is a Drag feels like coming together as a community. As a queer person who isn't out and who hasn't really found a community to be a part of, it feels so good to find a community that blends my love of science and queerness."

"[Participating] really helped me connect parts of my identity together and embrace who I am as a person...thank you 1,000 times."

For our performers, we leave a longer lasting impression as they realize they can *be* the role models they never had growing up, mentoring the next generation of presenters, and bringing their whole selves to work.

"I've had difficulty being taken seriously in scientific spaces, and through drag I've experienced acceptance in a way I never did in STEM."

"Queers, in all our glory, exist in science. And it can embrace us as much as we embrace it."

We've already seen organic replication in places like the UK and been invited to support shows around the world. Demand is so high that our shows book out within minutes of registration opening.

That's because Science is a Drag isn't just a queer show for queer people. The beauty of drag is it's so loud and sparkly and in your face that whether or not you consider yourself a 'science person,' you can't help but stop and pay attention.

Imagine my delight then, when our scrappy team of volunteers had the opportunity to pitch our project at an international science engagement competition in the Fall of 2022, vying for the prize of the Falling Walls' Science Breakthrough of the Year. Imagine my horror when they decided I should be the one to present. Don't get me wrong, I've delivered a fair number of public talks in my life – but never in drag. I'm the organizer. The one pulling strings behind-the-scenes to make our shows happen. I'm certainly not the star of the show.

Something about confronting the masculine parts of my identity and so publicly claiming labels for myself I'd never used prior sent me through an emotional rollercoaster, forcing me to confront my own internalized homophobia. Even living as an out queer woman for many years, I still had parts of my identity that I shied away from and felt shame over. They were fine for everyone else, but not for me. What would my parents think? My partner?

These questions and more raced through my brain in the months and weeks leading up to the presentation. It took weeks of therapy and many tears before I remembered that the whole point of this project is to remind people, including myself, that we are enough exactly as we are. That it's okay to take up space, to experiment. And so, newly emboldened, I got to work hashing out a script and hunting down an outfit.

I start the presentation as 'Rich,' an arrogant white man in a three-piece blue-gray suit which was donated by a lovely woman from life before her transition. She explained it had negative connotations and was all too happy to see it have a queer rebirth on an international stage. As I explain the concept of Science is a Drag to the audience, I slowly peel off layers of clothing, eventually unraveling a long black skirt which I'd bundled around my chest. As I wrap up, I reintroduce myself as me, as Carrie, shaking out my long gray/brown hair and presenting in a more feminine light. Happily, the emotional labour paid off, and I'm proud to share that we received a Special Award for Inclusive Science Engagement.

I don't purport for a second that our show is the antidote to queer exclusion in STEM. But with an increase in hateful discourse and violence towards queer communities, Science is a Drag is a sanctuary rooted in the joy of learning science through the powerful art of drag. When loud voices decry critical thinking and the world feels, frankly, like a dumpster fire, I firmly believe art, creativity, and good humour will pave the way to healing and inclusion.

Carrie Boyce

Carrie Boyce is the Executive Director of the Royal Canadian Institute for Science (RCIScience), Canada's oldest scientific society that's been connecting Canadians with science since 1849. With over a decade's experience working in the field of science communication and public engagement, it's fair to say Carrie's become a Jack of all trades, master of some.

Originally from Northern Ireland, Carrie moved to Cambridge, England, to pursue a degree in Biological Natural Sciences at the University of Cambridge, before working for the University and organisations like the Royal Society of Chemistry and Cancer Research UK.

Eager for life's next adventure, she moved to Canada in 2017 and has been happily working with RCIScience (and drinking maple syrup) ever since. In 2022, she received the Special Award for Inclusive Science Engagement at the international Falling Walls Breakthrough of the Year competition for her work co-producing Science is a Drag, an initiative which aims to shine a light on 2SLGBTQIA+ inclusion in STEM.

If you're interested in learning more about Science is a Drag and the upcoming shows, follow the adventure on Instagram @scienceisadrag or visit scienceisadrag.com. Who knows? Maybe one day you'll walk by a bar to find a pop up in a city near you.

Connect with Carrie on LinkedIn or via RCIScience, @RCIScience on social media or email infomation@rciscience.ca.

CHAPTER 2

The Evolution of a Biologist: From Studying Life to Reigniting It

Cathy Derksen

"If we want people to fully show up, to bring their whole selves including their unarmored, whole hearts—so that we can innovate, solve problems, and serve people—we have to be vigilant about creating a culture in which people feel safe, seen, heard, and respected." ~Dare to Lead by Dr. Brené Brown

Life can take us on some very unexpected journeys. If you've met me in the past decade, you will be surprised to learn that my original education and career was in biology/genetics. My work today is focused on serving a global community of women through my business, Inspired Tenacity. I support women in creating a life that inspires them through group coaching and exciting projects like this collaborative book.

So many people have asked me how and why I made the enormous shifts in my career. Why did I walk away from the field of genetics that I loved? The story of my career transition was impacted greatly by the workplace culture I experienced and my lack of support and knowledge around creating change at that time in my career.

To put this in context, I've been a biologist since the age of three. Those are my earliest memories of pursuing my curiosity and love for the magic of life. My parents learned to accept the ever-evolving zoo in my bedroom that would vary depending on where we lived at the time. My family moved several times when as I was a child, as far south as Nebraska and as far north as the North West Territories of Canada. Everywhere we went I created a new collection of the local creatures from snails and caterpillars to snakes, frogs, lizards, etc. Of course, I also had the usual pets like dogs, cats, rabbits, and fish. I loved them all and was curious to learn as much as I could from my experience with them.

Needless to say, it didn't surprise anyone when my education and recreational activities continued to focus on life sciences and spending time in nature. When I went to university in the 1980s, the world of genetics was advancing quickly. I discovered a whole new way of exploring deep into the magic of living things, and I jumped in with both feet. The field of genetics blossomed into new levels of discovery with genomics, proteomics, metagenomics, and the 'omics' collections continued to expand. It was a very exciting time to be involved in the field. The Human Genome Project led to an explosion of growth with new developments in technology and enormous expansion of our understanding of these building blocks of life.

I had discovered a career path which would give me the opportunity to continue my love of biology and absorb myself in an ever-evolving opportunity for lifelong learning. For twenty-five years, I worked in the field of Medical Genetics/Genomics in hospital labs, public education, professional certification and exam development, and conferences. I loved the fields of biology, genetics/genomics and studying the magic of life. Even while I was on maternity-leave, I continued to take courses, attend review clinics, and do research projects to continue my education and involvement in the community. It's been over a decade since I left this work behind but, to this day, I stay in touch with peers and the current news in the field.

You are probably asking yourself, what pushed me out of the field that I love so dearly? Many chapters in this book will describe cases of gender bias and discrimination that women were subjected to by the men and cultures in their work settings. My situation was the reverse. A vast majority of the people I worked with, men and women, were wonderful, supportive, cooperative team players. Unfortunately, it only takes a small group of negative, abusive people to shift the atmosphere for everyone.

One source of the trouble in my workplace was created by peers. A couple of my female co-workers made a game of bullying and sabotaging the work of others. I'm not just referring to small petty issues. They would create situations that led to others being disciplined and written up for incidents that had nothing to do with the accused. These workplace bullies were clever in disguising their involvement. They made a point of being positive and outgoing to the supervisors, so they were able to have any complaints against them dismissed. This destructive bullying went on for years.

The second source of our negative workplace situation came from above. The doctor in charge of our group was very set on maintaining a clear hierarchy in our workplace. Her need to maintain a position of top authority was forced on us in the form of constant belittling, micro-managing, and blatant blocking of our opportunities to build on our skills and education. She was suspicious of anyone who asked questions that might increase their knowledge and she made it almost impossible for us to attend courses or events to learn and grow.

When you take a group of highly educated, outgoing, creative, curious, keen people and tell them to just sit down, shut up and focus only on the work that is in front of their nose, you create a group of very frustrated, insulted, individuals. A few of our team members left to get away from this negative, abusive situation. Many of my co-workers stayed, despite the toll this was having on their mental and physical health. They felt they had no options, and they needed the pay cheque. I put up with this workplace situation for five years. At the time, I didn't feel I had any support to address the situations and saw no options besides leaving completely. My field of work was very specialized, so there were few alternative employers. Because of this, leaving my job also meant leaving my career.

Faced with the enormous decision to leave my career and stable job, I knew I needed to make some big changes in many aspects of my life. I went through a period of self-discovery and review of my skills, interests and passions. I recognized that I preferred to serve people in-person and that my heart was called to focus on helping women in particular. This shifted my path and I refocused to pursue a whole new direction in my life.

After a decade of my personal evolution, following my calling to be of service and my need to support other women in following their dreams and passions, my work now focuses on helping women create a life they are excited to live. I support them in building the courage and community to create positive change in their life and in their careers.

Looking back at that time in my life, when I left behind a twenty-five-year career in a field that I loved, I see how the landscape of working in science and technology has evolved. I'm not saying that workplace cultures like the one I was subjected to don't exist anymore, I'm saying that I see an overall shift in the landscape of women working in STEM (Science, Tech, Engineering and Math). As the discriminatory culture in many workplaces became exposed, we recognized the negative impact it was having on women and many women spoke out. A passionate group of women have taken the lead to speak for those that don't feel safe to do so. It is no longer an accepted norm for women to be paid a lower salary than men while performing the same job. It is not acceptable for women to be held back and skipped over for promotions based on stereotypes and assumptions around their career and family plans. We have made significant progress, but there is still a lot of work to do in order to reach parity.

Although I am no longer actively working in the field of biology, the work that I am doing to lift women's confidence and to create platforms for them to find their voices, to speak freely and openly, is supporting the shift in culture that is leveling the playing field in these careers and workplaces. I am supporting women who are feeling stuck and frustrated to reignite their life. By rediscovering their brilliance and their voice, women are stepping forward to become disruptors and catalysts for positive change.

I love to reflect back on this quote by Margaret Mead:

"Never doubt that a small group of thoughtful, committed people can change the world; indeed, it's the only thing that ever has."

By working together and collaborating in ways that build on the wisdom and reach of the group, women around the world are creating a new version of what is acceptable and supported in our careers and personal life. As you read through the chapters in this book, you will be amazed at the strides that have been made, and the wisdom that has been shared. This will allow us to build the future of careers in science, technology, engineering and math for women globally.

My goal in bringing this team of amazing women together in this anthology is to support them in elevating their voices. Sharing the stories of our triumphs, challenges and vision for the future can strengthen the community of women in STEM. We can inspire other women to be courageous in stepping into their higher goals and dreams for their career and personal life. We can collaborate, and support each other, in creating strategies for building an inclusive, supportive cultures.

I am so inspired by the women who have shared their stories in this book. I would like to dedicate this book to all of the women around the world who are dedicated to shifting the landscape of careers in science and technology. You are changing the lives of your peers as well as the next generation of women in STEM.

Cathy Derksen

Cathy Derksen is the founder and owner of Inspired Tenacity Global Solutions Inc. She is a Disruptor and Catalyst dedicated to improving the lives of the women in her community and around the world. Cathy helps women rediscover their brilliance, find their voice and create a life they love.

Cathy is an international speaker and eight time #1 bestselling author with stories that inspire the readers to take a leap of faith into reaching for their big goals. She has created a platform supporting women to share their own inspiring stories in books. With her all-in-one program, Cathy takes you from chapter concept to published bestselling author in a simple, exciting process. If you would like to become a bestselling author, connect with Cathy today.

A decade ago, Cathy transformed her career from working in Medical Genetics for twenty-five years to financial planning so that she could focus on helping women create personal success. Over the years, Cathy has followed her passion for learning and has become certified in counselling, success principles, and strategies for overcoming limiting beliefs and mindset. Her programs at Inspired Tenacity allow her to blend all of her skills to amplify the impact of her work.

Cathy has two children (twenty-eight and twenty-nine years old) and two fur-babies. She lives near Vancouver, Canada.

She enjoys spending time in nature, traveling, meeting new people, and connecting with her global community.

Connect with Cathy at https://inspiredtenacity.com.

CHAPTER 3

Journeys in STEM

Claire Skillen

But the whole of life is a detour, that's what makes it so fun; remember to take your eyes off the map from time to time though as the scenery is beautiful. A mentor of mine once shared this with me and I can't think of a better way to describe my career so far. Each experience along the way shapes and teaches us. I'll share a few of mine along with some thoughts for women in science, technology, engineering and math (STEM), leadership, and some of the pieces of wisdom that guide me and hope that some of it resonates with you. Interest in STEM can begin at a young age.

Since I was elementary school age, I found myself fascinated by how things worked be it nature, machines, or the stars. I remember my TV time being an equal split between my favorite cartoons and science and nature documentaries. You will still find a shelf full of Audubon field guides on my living room bookshelves at home. Of course, back in the '80s and '90s there were fewer examples of women in STEM. I, however, had the uncommon privilege of having two female cousins with engineering degrees as possibilities, so to me women in STEM were perhaps a more normal occurrence than both past and present statistics indicate.

I pursued a Bachelor of Science at the University of Victoria in physical geography with a minor in environmental studies and specialized in climatology and biogeography. The study of the natural and physical sciences fuelled my imagination and I found myself inspired by all the men and women committed to a continued understanding and protection of the world we live in.

Upon graduation, my adventures in STEM found me working as a fisheries observer doing catch monitoring on west coast fishing vessels off the coast of British Columbia; writing a management plan for Juan de Fuca provincial park; stream mapping Craigflower creek in Victoria BC; completing plant and animal transects and ecozone mapping for the Victoria Natural History Society; and a life-changing three plus month community development project in Costa Rica that included working with local biologists. All of these experiences showed me how much you can learn and the kind of difference you can make working in STEM. A couple of examples include the work we did with the Victoria Natural History Society, which provided the evidence necessary to convert the area studied into parkland; and the work with BC Parks on the management plan, which began to build my understanding of systems and the biopsychosocial, cultural, and spiritual elements of nature. I consider myself a systems thinker to this day.

Home from an enriching and perspective broadening experience in Costa Rica, I came back to a bit of a recession and needed to take a bit of a career detour. I found myself working on a leadership project with a BC Public Service, strategic human resources team – a lesson in leadership. A key takeaway was the importance of connecting and anchoring people to common vision and values system so that the people breathing life into the vision can weather the storms that come with the task.

I spent nearly four years with the women's and seniors' policy and employment and income assistance teams. These experiences impacted me profoundly as I learned the reality that women faced and still do in areas of gender-based violence, pay inequity, poverty, and access to services to name a few; a notably harsher reality experienced by BIPOC women, LGBTQI2S+ and women with disabilities.

A permanent opportunity moved me into the natural resources sector with the province of British Columbia in the oil and gas, and mining and minerals divisions. These roles brought my first opportunity to connect with Indigenous Nations. An elder in northern BC approached me at a community meeting and shared "you measure distance in kilometers, and we measure distance in memories." Fourteen or so years later, this stays with me as a reminder of the importance of connection to the land and how much there is to learn from and appreciate about Indigenous cultures. I had my first semiformal leadership opportunity at the BC Ministry of Energy Mines and Petroleum Resources as Chair of Innovation. I had the fortunate opportunity to work with an amazing team of heart-centred people that helped enhance a culture of innovation in the organization. One thing we all shared was a passion for being of service and having what I call brains without borders. I learned that passion, purpose, and imagination and can over time catalyze beautiful transformations; we were recognized by executive for our efforts. This experience also reminds me to persist when the shifts we hope to see seem to be taking their sweet time.

My first assignment as management at BC Hydro working in generation and transmission and distribution was a colorful one. My team of seventeen taught me how important it was to know my natural leadership styles and that I needed to adjust them as fluidly as possible to suit the situation, the team, and the person.

My time at Hydro was brief as there were mass layoffs and so I found myself embarking on another detour needing a map, a compass, and some advice. This time taught me the value of coaching and mentorship. Sometimes when times are tough, we have to borrow confidence and sight on occasion to help us navigate the trickier terrain on the life map. I think you will find that most if not all in this book have had a coach or a mentor at some point in their lives.

I have often said that new directions and opportunities find me. Had you asked me twelve years ago if I would be working in the technology sector, I probably would have looked at you quizzically saying that the thought never occurred to me and no I was not interested in programming, nor was I great at it. I am so thankful for all who are! After several months of job searching and the completion of an amazing leadership program with the Minerva Foundation of BC Women, I found myself on a formidable learning curve as I learned what it meant to function as a business analyst.

I spent close to seven years as a business analyst, which taught me that technology was about so much more than coding and that there are an infinite number of ways to be of service and make a difference even without a computer science, engineering, or mathematics degree. Please note, women of all walks, may you flood the gates and ranks of all three of these disciplines. The technology sector needs diversity in all its forms to build innovative and inclusive solutions that truly benefit broader communities.

This chapter of my life also taught me how much technology is really about people. In most of the pain point inventories I gathered for clients, I found that more than 75% could be resolved by working with, supporting, inspiring, and sometimes challenging people. I also learned how important it was to be strongly anchored in who I was and have a solid support network to weather the challenges that come with being a woman in STEM. Like a strongly rooted tree, you can bask in the warm winds, grow from the rains, and bend and flex in the storms. You may lose a branch or few, but these can regrow. If you find, however, that despite your strong roots your branches are dying and the roots are becoming shaky, then uproot and move to where you will thrive.

I currently work as a senior organizational change lead through my company Ceascape Solutions Inc. and have just finished four years as Board Chair for Island Women in Science and Technology (iWIST). Serving on the board and being part of this community was a true privilege. Some key takeaways from both change leadership and my time on the board: 1) a difference is made one person, one conversation, one act of service at time; 2) some of the seeds of change we plant are not for us but for the next generation to shelter under; 3) leadership is a privilege and a combination of caretaker, strategist, protector, advocate, catalyst and sanctuary in my experience so far (I constantly strive to show up this way); 4) stay curious and be brave enough to unpack your biases. Yes, we all have them. Uncovering, acknowledging, and owning our biases is in part what allows us to better connect, amplify, stand with, and celebrate diversity; communities are richer, healthier, more innovative, more successful, and more beautiful as a result.

Another takeaway is the power of language. One word can truly make the difference between someone exploring STEM or not, applying on a job or not, or feeling like they have permission to show up as their whole selves or not. The latter is a critical one, which is why I encourage a language shift away from the word "fit." Fit can imply a surrender of self right at the outset and I think you may agree that when people are able to show up as their whole selves, accepted and celebrated for who they are, they are healthier, happier more productive, and capable of amazing things. One word can make that much of a difference.

I have often wondered what life would have been like had I chosen a different path. I thought about pursuing engineering like my cousins, but at seventeen years old, having just graduated from high school, I found myself intimidated by math and chose not to pursue it. I would learn about twenty-five years later that engineering has little to do with math. If only I had someone to challenge my misconception and my fear of math, how different my path could have been. I am curious how many paths for young women might have been different had there been someone to help identify and challenge limiting beliefs and lend some confidence in these areas until it became self-sustaining. If you have a chance to do this for someone, do it. This is what organizations like Island Women in Science and Technology and numerous other organizations that support women and girls in STEM do. The power of community creates belonging versus fit, living, walking possibilities – representation, fun, learning, a place to laugh when things are good and a place to lean when things become challenging. We are all community builders.

Thanks for joining me on this journey through the ages. I'll finish off with a few of my favorite pieces of wisdom: 1) it's hard to catch what the universe has to throw you when your hands are full, so don't be afraid to empty them; 2) you have figured out everything life has thrown your way so far, so whatever you want to go for next I think you are a pretty good bet and remember to slow down and be in the moment (I am speaking in particular to the A type personalities out there); 3) life passes all too quickly, so take the time to reflect and celebrate all that you are and have accomplished! Best wishes!

Claire Skillen

Claire Skillen is Founder and President of Ceascape Solutions Inc., where she contracts as a senior change lead for the province of British Columbia and past Board Chair for Island Women in Science and Technology (iWIST). She is a passionate community builder, award winning leader and supporter and advocate for women and girls in STEM. When not at work or in the STEM community, you can find her hiking the trails, surfing her outrigger canoe, cold water swimming taking photos, spending time with those closest to her, listening to some great reads on audible, and catching some beautiful sunrises near Clover Point.

Connect with Claire at www.linkedin.com/in/claireskillenmbabsc.

CHAPTER 4

Coloring Outside the Lines

Diya Wynn

Do you remember coloring as a child?

We typically start off coloring quite young. It's a pastime we enjoy. When we first start, we are allowed the freedom to create in whatever way our hand graces the page, with whatever mix of colors we choose to form the masterpiece we've envisioned. We were not constrained to the lines on the page. There were no boundaries. At some point, there was a shift and we were given guidance and instruction to keep within the lines, maintain structure, and choose appropriate colors. With age, came a value on structure and order. The better picture was the one that followed those rules.

I remember having a Crayola 64 set with its own sharpener. I enjoyed coloring and felt I was quite good at it because I was structured, neat and precise. I took great effort to ensure I followed the rules and applied great care in the colors I chose. After all, it mattered in how my art would be perceived and what would be valued. As children, we don't start off with that sort of precision and structure. Coloring is a learned, foundational skill that evolves through stages of a child's early development; hand-eye coordination, fine-motor movement, dexterity. From infant, toddler, preschooler, to school-aged, our coloring improves and our skill develops to where we have the self-control, organization, orientation to space, awareness of boundaries and focus to color inside the lines.

I settled into this idea of coloring outside of the lines in an early briefing for a speaking engagement with an organization in Australia. We were meeting for the first time and exploring what would work well for the audience. They wanted to hear about what inspired me to explore a career in technology and how that influences the work I do today. Why coloring? What does coloring outside the lines mean? How could this inspire folks at various stages of their career?

In the third grade I received a computer as an award for having the highest reading and math scores on the standardized tests. Getting a home computer was a big deal! Today we have so much computing power in our pockets, but that's not how it was back then. Most people didn't have computers in their homes, and it certainly was not common in the homes in my community.

The computer I was gifted was pretty basic. The Commodore VIC-20 had 5 Kilobytes of memory and half of it was used by the operating system. We have more computing power in our pockets than I did on my computer. It was basic in black and white… or brown and orange. I think the most complicated and memorable thing was this game called pong with a rectangle block on each side of the screen and square "ball" that bounced back and forth. Despite the fact that the computer was primitive, it fascinated me. And unbeknownst to me at the time, it opened up a world for me unlike anything I had been exposed to or frankly was aware of before.

None of the fairy tales I watched had women with jobs. None of the books I read told stories of women working in computers or science. They didn't tell stories of women in technology careers in any of my school studies. It was not something I saw in the women around me either. What gave me the idea of being a computer engineer? Who knows! But at eight years old, that is exactly what I decided. I was going to be a computer engineer.

As a result of getting that computer, I was exposed to technology and it opened up a new realm of possibility for me, and I was determined. I didn't know much about technology prior to that other than maybe the black-and-white tv I had with turn knobs. Getting a computer as an award was a big deal. I was too young to understand the gravity of my decision then, but I embraced it. It fueled my curiosity and passion that led me to matriculate at Spelman College, a historically black college in Atlanta, Georgia, to study Computer Science. This was despite being told that graduates from the private preparatory school in Connecticut I attended don't go to HBCUs. I was the first. From that third-grade exposure, I set my mind and my interest in technology, and I've never wavered.

Coloring inside the lines would have been to follow in the footsteps of my grandmother or mother and become a teacher, or to follow in the traditional roles they portrayed in media for a woman – a homemaker or maybe a nurse. Isn't that what I was taught?

Until the late 1960s, Jim Crow laws dictated when, where and how blacks could exist in the world. They constrained opportunity, access, and dignity. By the time I was growing up, these oppressive laws weren't active, but were and still are systemic barriers and, even hidden, stereotypes that inform the expectations of others about what we can do, what we're capable of, and what success we can achieve. I wasn't determined not to be limited by that. My faith wouldn't let me be diminished to a reality beneath the vision I had of myself and I believed God purposed for me.

My story may not be common. Most people change their declared major in college multiple times, let alone keep the career focus they declared as a child. Let's attribute this foundation of focus to coloring inside the lines. Early childhood philosophy also finds that the level of focus and concentration from coloring helps children complete many other tasks and can foster their confidence.

It was the exposure to technology at an early age and education that changed the trajectory of my life, and coloring that gave me the confidence to step out of the box and do the unconventional. Coloring outside the lines requires an awareness of boundaries, space, order, and structure coupled with an intentionality to create new boundaries and new order – to think and act in a way that does not conform to the norm. It hasn't always been neat and pretty, but it's resulted in a beautiful picture in the midst of all of the mess. That's a great reflection of what my life has been like.

Imagine stepping into a career and an environment that didn't look like you, wasn't made for you, and never intended to be for you.

Now, in my current role, I have the opportunity to influence and shape the ideas of others – major companies across many different industries globally as they build computer systems and products using artificial intelligence – with the intent of driving more fair, inclusive systems that deliver more equitable outcomes that benefit us all.

Like other technological advancements, AI is transforming the way we live, work, and play. It is creating new jobs, driving new income streams, fostering stability in markets, and establishing avenues of access to resources....

It's opening doors of unseen opportunity. We want... need those opportunities to be open for us all.

I lead discussions globally to get companies that are using artificial intelligence in their products for business to take intentional action to uncover potential unintended impact and mitigate risks that often disproportionately affect the underrepresented, undeserved, and marginalized.

This is coloring outside the lines because most technology is built by a majority population, white men. As a technologist, I understand the promise that technology brings – the great things we see possible with every advancement. But with great power there is great responsibility. If companies are not intentional about how they design, build and use technology, it will continue to be *done to us,* not *for us*. Leaving the underrepresented out of the equation has huge implications.

This presence-absence is not only true for women but for people of color and other less visible communities of people. This gap has huge implications on technology and data that is driving the future.

We have the responsibility NOW to create space, open doors, and foster environments of safety, peace, belonging and inclusion that they will live into. Are using what we have to drive equity not just equality? We can't expect these young women – the next generation even – to take the reins of a world where we've not been good stewards. What are we leaving for them?

We (when I say we, it's a collective 'we' that requires the involvement of government, public and private sectors, industry and organizations) must be determined to address systemic and institutional barriers that perpetuate discrimination against and keep us all from being seen and treated as equal and being afforded opportunity. And that cannot be done by women or people from marginalized communities alone. It also requires the active work and partnership of men and others who find themselves in positions of the majority.

It's taken a little bit more to put it all into focus and see the beauty that is there. But it's been great and everything that I've experienced along the way has helped to bring me to this point in my career.

My exposure and career in Technology was coloring outside the lines. It gave me a pathway out of the hood and into a life full of defining moments where I've had to resist the urge to stay in the lines. To see value in what I bring to the world. To be seen, heard and lean into opportunities to create a better future for me and you, and hopefully generations to come. I, too, hope for you to find the courage to take your crayon and color outside the lines.

Diya Wynn

Diya Wynn has worked in technology for more than 25 years building teams and developing products in early and growth-stage companies. She has a passion for developing current and emerging leaders; promoting STEM to the underrepresented; and driving diversity, equity and inclusion (DEI). After spending years in technology-focused roles and separately pursuing passion projects in DEI, she now has a role marrying the best of both worlds as a global lead for customer engagement in Responsible AI. In this capacity, Diya leads a team and advocates for intentional action to uncover potential unintended impacts, mitigate risks and ensure inclusive and responsible development, deployment and use of AI/ML systems. In addition to her customer focus, she meets with legislators and regulators globally to influence imminent regulation and policy on AI.

Additionally, Diya is an inspirational speaker; has been awarded and recognized in the industry for her work in Responsible AI; serves on non-profit boards; mentors through multiple organizations including her alma mater, Spelman College; guest lectures regularly on responsible and inclusive technology.

Diya studied Computer Science at Spelman College, the Management of Technology at New York University, and AI & Ethics at Harvard University Professional School and MIT Sloan School of Management.

Connect with Diya at diya@diyakwynn.com

CHAPTER 5

My Journey to Sustainability

Dr. Doris Hiam-Galvez

Many of you may think that there is no hope for the future. I think otherwise and I offer a message of hope and practical steps on realizing a better future.

After thirty years of engineering experience creating new markets with innovative products/services/technologies on four continents, I developed a structured method to enable the creation of a future sustainable society.

Trained as a PhD in metallurgy and leading research teams for manufacturing companies supplying to the automotive and aerospace markets has equipped me to anticipate the future and invest in research for long-term results.

Strategic materials are vital to stopping climate change and, therefore, the extractive industry has a chance to become a catalyst for a sustainable future.

Immersed in Nature

Growing up immersed in nature, breathing clean air, singing with the birds by the riverbanks was one of the most beautiful memories I have from my childhood in Peru. Exploring the rich biodiversity from the back of a horse were adventures I will never forget. Here is where I learned that we are never alone and when we are connected to nature!

I was attracted to rocks by their beauty, and I collected rooms full of them. I wanted to be a miner, but I could not study mining in Peru since the belief was that a woman entering a mine was bad luck. Therefore, I studied chemical engineering to leave the door open for anything that would inspire me later, then quickly focused on metallurgy. My first internship and my first job were at major Copper & Zinc metallurgical complexes.

Growing up was hard, but being immersed in nature made every setback an opportunity to learn and be ready for the next challenge.

Life is just overcoming obstacles and the challenges we confront are just small battles to prepare us for the real war.

Around the World

Peru is a major producer of Zinc in the world. Zinc has an affinity for Iron and is the reason for the whole field of galvanizing where the zinc sacrifices itself to protect the iron from rusting. My name happens to be Galvez, so maybe I was destined to work on galvanizing. I went to Belgium to do my PhD to study the reaction of Zinc with Iron with a world-renowned professor, but first I had to catch up with metallurgy. I was studying a new subject in a new language in a new world. I was in lectures, but I could not understand the language or the subject. I was close to giving up and returning home, but I built up the courage to continue and after the first year ended and I was ready to work on my PhD, the Professor died!

If there was a war, this was it. Was I strong enough to overcome this obstacle?

I found two other experts, one in Germany and one in Canada, to review my work. The PhD experience was hard all the way, but the support from the university, the industry, and the advisors helped me to overcome these difficulties. The support from industry was key to develop a new product. This added value to my doctoral thesis.

After finishing my PhD, I joined a steel company in the USA to work on galvanizing.

Even though I did not know how to drive a car, I managed to obtain a driver's license, bought a new car and become a public danger. My English was poor and I could not communicate well. One night I visited the galvanizing plant. Luckily, I arrived at my destination where I heard loud voices in Spanish. I was so happy to hear my language. The operators were big and a bit scary, but they were extremely helpful. Because of their support, we were able to develop a new type of galvanized steel the company needed to capture a new market in the automotive industry.

A year later, I married the big boss. I remember meeting John when I visited the company for the job interview. When I entered his office, I will never forget what I saw in his blue eyes. I saw my past and I felt that we had always known each other. I was happy to have found an old friend. We both loved nature and the first time we went out together it was to a park by the Great Lakes during Easter. During that outing, we realized we had the same interests and values. John has been and still is a great mentor, inspiring me to achieve what seems impossible.

A year after our marriage, we were blessed with the arrival of a most incredible human being that immensely enriched our lives. Our beloved son changed our lives that became extraordinarily enriched. There were many obstacles to overcome trying to balance work and family without having any family support in a new country, but the deep love connection between us and nature inspired us to overcome every battle we encountered. These were the times of true love for each other, for nature and everyone. It was the best time of my life.

Sometime later, I decided to explore other industries and arrived in aluminum manufacturing. Although I changed industries, I was expanding my responsibilities from research to all aspects of the business. My career in manufacturing went from research leader to director and then to chief technology officer for a major global corporation.

Inspired by the Sun

The second part of my career started in Canada. I joined a global engineering company providing services to the metals, energy, and infrastructure sectors. Very soon after I joined the company, I went to Australia to expand our business by leading an acquisition. After a successful integration of the newly acquired company, I went to South America to lead the expansion of the business in that region. Although there was a recession, and the market was full, I was able to build a comprehensive operation to provide new services not available in the region. The key was to have attracted the brightest young professionals and create the environment to unleash their highest potential. These innovative services were expanded globally and became the basis for core services for the company. This young team made history and they continue shining light into the world.

After this experience, I went to London (UK) to lead our European operations in a mature market with readily available innovation. We managed to penetrate Germany, France, Sweden, etc. by combining multiple disciplines to solve environmental challenges and overall manufacturing process efficiencies. For example, we embarked on a path to rethinking mining and piloted a novel approach of extracting minerals from the ground without making a hole or building access but, rather, processing the ore on site. We would use geoscience to better understand the characteristics of the ore body to design solutions to minimize environmental impact. During this period, I had the freedom to do what is best for our clients and the company. This freedom drew the best out of me and the people I was fortunate to work with.

For the last ten years, I have been taking a lot more risk and approaching challenges in a wholistic way by combining more disciplines and figuring out how to achieve significant results. This next phase was inspired by my current boss who is a man with a heart in the right place. His genuine passion for sustainability inspired me to build courage to develop a new way of doing business.

Travelling around the world and working with investors in the natural resource extraction industry inspired me to develop a new way of doing business that leaves behind a prosperous economy with an improved environment. The objective is to diversify the regional economy during the time of boom of the extractive industry so that some level of economic activity continues after the natural resources extraction has ended. Primarily the focus is on the extractive industry, but the concept applies to any region. It is called "Designing Sustainable Prosperity" (DSP). DSP is a method that integrates multiple disciplines to visualize the hidden potential of a region (human and natural resources) and uses this information to inspire a process where key parties determine long-lasting innovative enterprises (products, services, and technologies) ideally suited to a region to compete in the world market. It is a co-design process with the key players from beginning to end. To be sustainable, locals must lead these enterprises, so it starts by empowering locals and positioning them at their level of strength and from this position the region's unique solutions are determined. From the depths of the hearts of the people and the depths of the earth a breakthrough occurs and unique solutions for the region are designed. The method uses geoscience to unearth the potential of the region and connection to nature to unleash the human potential. It starts with one champion. Would you like to be the champion for your region? Observe what is driving your regional economy and what other things could be done there. A book will be published in 2023 detailing this method so you can use this approach for your region.

Once the focus for the region is determined, the local education may need to be adapted to provide the skills and tools to support the diverse economy for the region. For this, I am working with a world-renowned perception expert to transform the way students learn, expanding their perception so they can think innovatively. We are piloting this new program in Canada in 2023 where 100 teachers and their students will be the "scientists." This is to infuse the love of learning and the love of science. After this pilot, the program will continue being developed for all levels of schooling including university education and available to the world. This information is invaluable to caring educators in Canada working with 8, 9 and 10 grades and is crucial for them and their classes to take part in this first-of-its-kind week experience in 2023. Teachers from anywhere in the world will also benefit for the next phase that will focus on teacher development.

My life quest has been to ignite people's highest potential adapting many techniques to unleash their potential and for them to do truly innovative work. Some of these techniques are co-creating with nature, doing art and music without any training. I had to do it myself as well and this pushed me to the edge. I composed a song ("Up with the Sun") and sang and recorded it professionally without having music or vocal training. The secret is "Letting Go." You can do this, too! I did this to encourage all of us to get up with the sun, open our hearts, and receive abundance in all aspects of our lives.

What drives everything I do is love for the planet, and now I am ready to contribute significantly to reducing global warming. This is my biggest ambition so far and requires collaboration and partnership between all the players in this objective.

One aspect I will focus on is studying further the sun as a direct energy source to desalinate sea water and power other processes. For example, the trees capture the sunlight for photosynthesis and produce the most beautiful and delicious nectars we consume to live. Can we do this as well?

Dr. Doris Hiam-Galvez

Dr. Doris Hiam-Galvez is a Senior Advisor at Hatch. She is also Board Director for PDAC & Chair for CIMBC22.

She is a doctor in Metallurgy from the University of Liege, Belgium. She has broad senior level experience in technical organizations and in developing new business globally focused on value creation.

For 16 years she has been creating new and innovative business to expand Hatch in new regions (Australia, South America, Europe, and North America). She managed Hatch Peru until 2012 and Hatch Europe until 2017.

Before joining Hatch, she was Chief Technology Officer for Novelis, a major global metal manufacturer. She was also a Board Director of the Canada UK Chamber of Commerce in London.

Working with clients around the world who were struggling with sustainability inspired her to develop "Designing Sustainable Prosperity" (DSP), a new way of doing business which leaves behind a positive sustainable economy with an improved environment.

She is happily living in Vancouver with her beloved husband, John, and her son Kamir, PhD in immunology, lives in San Francisco. Another member of the family is Sooty, a cat who has a unique ability to make eye contact with humans to get what he needs.

Doris' hobby is immersing herself in nature. She looks for the sun every day to receive daily inspiration. She also uses art for inspiration. If you are working on sustainability or on any aspects of the sun, please connect to collaborate with Doris at https://www.linkedin.com/in/doris-hiam-galvez-bb94761.

CHAPTER 6

Female Empowerment: A 5-Pillar Superpower Methodology

Gayle Keller

I was not always outgoing. In grade school, I struggled because I wasn't popular. Girls were mean, so I became known as "the mute." Fearing judgment from others, I was afraid to speak out. I wish I had the superpower back then of not worrying about what others thought of me. It wasn't until high school that I blossomed into myself and found my voice through the speech team at Glenbard West High School in Illinois. Now, what inspires me to take leaps of faith into the unknown is the mere fact that I do not want to wake up one day and wonder what might have been if I did not try something new.

In front of those infamous lockers in high school is where my speech team friends and I practiced our speeches and scripts. It's where I gained the confidence to be myself, to be 100 percent, authentically me. Participating on the speech team taught me the perseverance and communications skills I still tap into today. In college, I discovered my leadership traits when I became the Vice President of the Indiana University Student Foundation ("IUSF") under the direction of the greatest networker and mentor I know, Curtis R. Simic, President Emeritus of the Indiana University Foundation. Mr. Simic instilled in me the value of philanthropy and has become a good friend and mentor.

Both the speech team and IUSF gave me my wings to fly, and my courage gave me the confidence to love my unique traits and set them free. Now that I am a mother of two girls, my wish for my beauties is for them to remain curious, laugh at themselves, have fun, never lose their voice in fear of judgment, be resilient, stand up for what they believe in, "find the beauty in the ashes," be respected and respectful, and love their unique traits. I hope that I can give them the character-building tools that will empower them to soar and leave a positive mark on this world.

My journey through childhood, adulthood, and motherhood has melded over the years. This book was birthed out of my experience as a woman in tech, as a woman in S.T.E.A.M. (Science, Technology, Engineering, Arts, and Mathematics), and as a participant in the continuous uphill battle women have to face to thrive in male-dominated industries and workplaces. Having researched the gender gaps in S.T.E.M. (Science, Technology, Engineering, and Mathematics), it became my mission to work with both girls and women—since the gap really begins at adolescence, both at home and at school. It's become a passion of mine to empower girls to find their superpowers and follow their dreams. In the same breath, I empower women to pay it forward and sponsor or mentor other qualified women rising in the ranks.

My personal transformation from the "A" in S.T.E.A.M. (art) into the "T" (technology) did not happen overnight. In my early twenties, I lived in San Francisco for "two years too short" where I moved from advertising into sales in the technology space. I tried for years to get into sales and kept getting rejected because I did not have sales experience.

Living in California, I fell in love with technology. I cannot fix a broken computer, but I can empower customers to digitally transform by delivering outcomes that solve their business challenges. I fell in love with the fast-paced environment, people, and solutions that enhance our world. My first job in sales was selling print advertising in a technology magazine. My manager at the time took a risk on hiring me, since I had no direct sales experience. I did not hide that fact. Instead, I tapped into my youth jobs where I worked pharmacy retail and in the county law library, highlighting the skills learned in these previous jobs. I also brought my other skills to the forefront, such as being a strategic thinker, someone who always brings a solution to the problem, respects my peers, and has an inclusive nature.

Long-story short, I landed the job and have been in the technology space ever since. Even though I cannot troubleshoot a computer to save my life, I can use my sales, marketing, and relatability skills to enhance the technology industry.

One of my biggest challenges—and opportunities—I've faced over the years is being the only female in the room. I've been spoken over, spoken down to, told I am too tenacious, told I have an attitude, and not been given a proper chance after proving myself. And by the way, I do have an attitude when I am gaslighted or disrespected. It's my first line of defense. I do not have room in my life for negative people who distract me from my goals. But what got me through these tough times was my superpowers—especially courage after I gained the clarity I needed to take my leaps of faith. For me, having courage means trying something new, even if it doesn't work out. I do not want to wake up one day and wonder what might have been. Even if the leap I take doesn't pan out, my mother taught me to always find a silver lining in life. So, even if it doesn't work out as planned (which happens most of the time!), something positive will come from the courageous leap. It may take months or years to figure out what that something is, but it will come, and I continue to keep the faith. Because I struggled, it became my mission to empower other women to shine in the industries working to achieve gender parity. We've got a long way to go to parity, but just like how an elephant eats, we need to tackle our dreams (and challenges) one bite at a time. That's how I empower both women and girls in S.T.E.A.M. when it comes to networking, building their personal brand, and taking risks.

Through my firsthand experiences as a WiSTEAM (woman in S.T.E.A.M.), I sought the best path forward to overcome obstacles and turn challenges into opportunities—while staying engaged, asking questions, and being a good listener. When I began my career over twenty years ago, my work ethic, determination, and drive empowered me to learn, grow, and travel the world, submerging myself in other cultures. These experiences made me curious about empowering others to follow their career aspirations, while overcoming challenging situations in the workplace.

In the advisory work I do, coupled with my firsthand experience and Theodora doll line, my wish is to give a voice to the women and girls who struggle with indecision. My hope is that those females see their unique characteristics as game-changing superpowers that will aid them in taking calculated risks in their careers to elevate, expand, and/or reinvent their professional lives.

As a society, we are great at pointing fingers when something is not working to our advantage and rarely look in the mirror to see how we can improve upon ourselves. This book provides situations and models for how we address issues head-on while remaining appreciated and respected.

When it comes to reinventing my professional life and taking calculated risks, I've reflected on, analyzed, and assessed the processes and practices I've followed to formulate a methodology that will lead other women to build the life and career of their dreams. In my presentations and workplace training, I teach calculated risk-taking based on the five traits that I myself practice to elevate, expand, and empower my career:

1. Decisiveness
2. Courage
3. Clarity
4. Confidence
5. Balance

Personally, I've learned that my superpower is #2, Courage. My international bestseller, *Full S.T.E.A.M. Ahead: Triumphant Tales for Working Women to Overcome Adversity, Fear, and Self-Doubt*, reflects not only my journey as a woman in tech, but tells the stories of other women who have struggled and persevered to become their best selves. The women depicted in my book illuminate the common issues women face in the workplace such as lack of confidence, work and family conflicts, competition from colleagues, workplace bullying, sexual harassment, and mental health issues. I learn best through storytelling, and what better way to share with the world how to gain these superpowers than through nonfiction stories (which include some of my own!). Each Theodora represents women who are strong, intelligent, and successful in predominantly male S.T.E.A.M. careers.

My work includes the personas of five Theodoras who have defined their key pillar, superpower, and profession, which has led them to their desired success by outlining and following their blueprints on how to improve in each of the depicted scenarios.

- Silla™ believes "Clarity" empowers her to be sensational in Science
- Tyriqa™ believes "Decisiveness" empowers her to be tenacious in Technology
- Elaine™ believes "Confidence" empowers her to be exquisite in Engineering

- Antonia™ believes "Courage" empowers her to be assertive in Arts
- Maalika™ believes "Balance" empowers her to be motivated in Mathematics

Because the world still thirsts for gender parity in S.T.E.M., and because I was an "A" that transitioned into "T," it is imperative that I share my story of a journalist turned advertising professional turned woman in tech. If I can come from an Arts background, and make a positive splash in Technology, other women can too. "If we see it, we can become it." If we don't see other women in leadership roles in S.T.E.A.M., women will get discouraged from majoring in S.T.E.M. fields and dropout. Or, on the flip side, these women will enter the working world with their S.T.E.M. degree and S.T.E.M. jobs, fail to see other females climbing the ranks, and switch professions or leave the S.T.E.M. fields altogether. Therefore, my second wish is that my story empowers others to stay the course and follow their dreams. And I challenge women who have been successful in these fields to pay it forward by mentoring and sponsoring qualified women looking to rise in the ranks.

Did I struggle to be successful in tech? Absolutely. Did I seek mentors and sponsors to compliment my aspirations to empower my roadmap(s) to success? Absolutely. However, they did not just fall into my path. I had to seek them out by networking. That's where my courageous superpower came into play. Did I go for roles and not get them? Yes. Courage gave me the confidence to take a risk. And when I failed, I "failed forward," learning from my defeat and tackling something else new and challenging at every turn. I knew I had a network and tribe supporting me and cheering me on for trying, whether I won or lost. I tapped into my other superpowers of clarity and decisiveness, always in search of balance in an ever-changing world, where both work and life is all about integration.

After extensive research, I stumbled upon the fact that it is not solely the job of companies to solve the problem of gender inequality in the marketplace. It begins with the community we surround our children with—from family to schools. Therefore, it is our duty to expose children to all S.T.E.A.M. activities—just like we do with exposing a variety of sports to our children to see which ones they like the most—from childhood to school and beyond.

The prequel to my adult book embodies the same adult superpower from a young age to tackle life's challenges with grace, kindness, and respect:

- Silla™ believes "Clarity" empowers her to be sensational in Science activities
- Tyriqa™ believes "Decisiveness" empowers her to be tenacious in Technology activities
- Elaine™ believes "Confidence" empowers her to be exquisite in Engineering activities
- Antonia™ believes "Courage" empowers her to be assertive in Arts activities
- Maalika™ believes "Balance" empowers her to be motivated in Mathematics activities

In addition to my children's book, I am working on a children's toy line to empower girls and let their superpowers shine. My wish for you is that you recognize yourself within them, explore your future, and find solutions that will work for situations you encounter in your profession.

To Your Inner Theodora,
Gayle Keller

Gayle Keller

Gayle Keller is a former Microsoft sales executive who spent the last 20+ years in tech and felt called to empower females by contributing to closing the gender gap in the S.T.E.A.M. fields after seeing firsthand the gender biases that still exist in the workplace (coupled with the births of her two daughters). Gayle empowers women--and girls in S.T.E.A.M.--to take calculated risks and follow their dreams through her five-pillar superpower methodology highlighted in her international bestseller *Full S.T.E.A.M. Ahead: Triumphant Tales for Working Women to Overcome Adversity, Fear, and Self-Doubt.*

Connect with Gayle at https://gaylekeller.org.

CHAPTER 7

Duckling

Hana Galal

It pained my parents to see me disappear into crowds, climb up the side of playground towers, and run towards wild ducks and baboons. Yet, they still encouraged my desire to explore the world around me and, with their constant support, I flourished. My family always made me feel like I could do anything. I painted, I wrote, I got involved in sports, and learned languages. With age, I have become slightly more cautious, but I continue to seek adventure. Every experience, and every time I travel, I gain a new understanding of what I want from life. My childhood urge to wander off and explore has translated into a happy life enriched by unique research opportunities, new communities, and lasting relationships.

Wanderlust

My father is Egyptian and my mother is German. They moved to Saudi Arabia in the '80s; I was born and raised there. I have lived in fifteen different homes, ten cities, four countries, over a twenty-year span. When people ask why I move, I don't quite know how to answer.

With an internationally-spread family, and friends who move around the world, I realize there is no particular city or country that holds all the meaningful components of my life. I don't plan for moves; they happen when I decide to follow an opportunity. At a young age, I would travel for school and sports, and I realized there was a community of people like me who understand how I feel, that constant underlying itch to explore; third-culture kids. While some of my whims make my decisions seem scattered, everything has led me to the life I want. Importantly, I have recognized that the risk of change is scary, but it is not as scary as feeling stuck.

Finding Direction

I studied in Germany at a university focused on international and interdisciplinary research. Our mascot was a duck, primarily because ducks could walk, fly, and swim – they adapt to most situations. I think that concept of exploration resonated with many of my classmates who felt the same wanderlust as I did. For my bachelors, I majored in Social Science, and minored in biology and conducted my research in happiness. Understanding how societies function and how people interact to improve their well-being only made sense by way of reflection on a natural science. A direct line that I drew through sociology, biology, and psychology, was the contribution of nature, community belonging, volunteering, and hopefulness to overall well-being. Happiness couldn't be explained through individualistic success, nor was it reflected entirely by the commonly taught Maslow's Hierarchy of Needs. There is no single path for finding meaning and happiness, but I was moved by the research that seemed to show clear patterns.

Environmental Research

After working for a year in my first job out of university, I was drawn back to research, and I couldn't let go of what I had learned about happiness. I wanted to be involved in the preservation of nature and connection to societal well-being. I applied for a Masters degree in resource Management and Environment. I jumped at the opportunity to move to Canada. There, I concentrated on urban planning, civil engineering, water management, and business. My research focus: how large-scale infrastructure projects could be environmentally sustainable and socially accepted. Pure science and research helped me understand details and building blocks of different fields. The insights created a powerful foundation, and I found I enjoyed working between fields, seeing how different research connects and supports each other. Ultimately, interdisciplinary work becomes valuable when it is connected to communities and applied to solve real environmental problems.

I started working for a Canadian airport shortly after my graduation. As a manager, I gained insights into environmental protection, infrastructure planning, government communications, and strategic planning. I had stumbled into a way to integrate different interests in an objectively positive way. I enriched my life outside of the office with courses in history and French, and lessons in painting and data science. After many years at an airport, I moved to consulting. I now work with airports globally, and further actualized my longing to travel and connect. I can use my expertise, still I approach situations accepting what I don't know and don't understand so that I listen well, and am always open to learning. One of the exciting things about working between fields is there is always a new way to connect research areas to each other by setting a common goal.

According to the Convention on International Civil Aviation in 1944, the purpose of the industry is to "create and preserve friendship and understanding among the nations and peoples of the world." This guiding statement continues to drive my interests in aviation and its ability to build communities. The future of aviation must be sustainable and environmentally conscious while delivering on its original mission.

Connecting with new communities

When projects are established or completed, I start to feel restless. Almost naturally, I feel drawn to places thousands of kilometers away. Bolstered by my curiosity, I drift away, and settle in a new ground.

Every move has taught me how to enjoy and appreciate being an outsider. I end up meeting other people who are in transition, between cities and countries all over the world. It can be difficult always being "new" and moving away from friends, but I have found that real friendships and feelings of community continue despite distance. With every move, I take true friendships with me.

Giving back

Volunteering allows me to connect with the community I am in and see my impact on an individual level. I have always gravitated towards teaching languages because I have felt the benefit of language skills in creating connections and giving people the feeling of belonging.

At the same time, I remember why my work is important and draw the common thread between my day-to-day projects and their outcomes to the people around me. Some projects can be so detail-oriented that we need to force ourselves to zoom out and think about the broader picture. When we aim to ensure our contribution is positive, it can change how we approach life and the meaning of a job.

Similar to the concept of sustainability, we need to think of our lives holistically; giving back to communities through volunteering, caring for animals and nature, and finding people to love and who love us back, gives life meaning.

Mentorship is particularly valuable in a field where women can feel like outsiders, no matter their role. I have had incredible men and women mentor me, listen to my concerns, and give me advice. Sometimes it was helpful to ensure I understood difficult situations appropriately, and I was considering the right risks at the right time. It has been important for me to mentor younger people entering aviation or sustainability to make sure that they know they belong and make a difference. I have received awards for my work with children and was nominated for the mentor of the year award for my support of students. Helping others fit in and connect will also feed into our own well-being. I enjoy my field of research, and the most meaningful component is the relationships I've built on my journey.

Hope for the Future

A large part of my consulting work is connected to developing improvements for future scenarios. In sustainability and infrastructure planning, we need to think about what communities might need now and in 25+ years. We need to consider how to protect access to safe aviation infrastructure and integrate potential technologies. We need to think about climate-change impacts and how to protect nature, people, and assets in our planning. In my mind, I associate forecasting straight back to a pursuit of happiness. When we try to prepare for the needs of the next generation, our perspective should be, how can the future be better. By protecting the well-being of others, we also are protecting ourselves.

When I started in my field, few universities offered the topics and interdisciplinary approach I was looking for. When I entered the workplace, few businesses had sustainability and planning roles. I have been part of research and contributed to conferences and cross-national groups committed to expanding the field. Now, there are sustainability leaders and executives that are concerned with protecting all components of our collective future – from aviation, to clothing, to food and hospitals. Ultimately, I think it will become a norm that most roles will need to tie back to responsibility to nature and societal well-being.

Women at work

Though I have had the opportunity to work with organizations who care about advancing women in the workplace, I have seen that in general, women need to push more for our place. Sometimes there is power in pushing back; sometimes it's powerful to know when it's time to leave a situation that no longer works for you. I have developed diversity and inclusion programs at organizations, so that everyone can feel that they belong.

It is important to know your value and advocate for it and not to waste your energy in the wrong places. I have learned many times over that we need to be careful not to set up our own barriers. There are enough real barriers in the world, and there are enough people who will say no. It happens to everyone that we play tapes of failures in the back of our thoughts, 'I couldn't or I shouldn't,' and we need to stop and ask ourselves why we don't trust ourselves to risk the unknown. Failure and criticism are part of progress. My mother's success and commitment to her work was a role model for me, particularly when I encountered difficulties. She made me feel like I had permission to ignore people who might say no, and carry on with what I saw was correct.

Conclusion

"Traveling – it gives you a home in a thousand strange places, then leaves you a stranger in your own land."

~Ibn Battuta (1304AD - 1369AD)

My dad used to lean down to kiss my forehead and call me Battuta. In Arabic, it means duckling, and it's also the name of the Maghreb's most prolific historical explorer.

My life has involved pushing through many different boundaries, from research fields, to countries, and social expectations. I continue to learn more about what contributes to my own well-being in my work and in my private life. The protection of home and community have become constant priorities for me. I have had the luxury to redefine 'home.' For me, home has become where my family and loved ones are. So, I'll keep exploring, because I always have somewhere to go, and I always have somewhere to return.

Hana Galal

Hana Galal is a Sustainability and Strategy consulting lead in North America for a specialized aviation consulting firm. She supports clients all over the world in sustainability, infrastructure planning, and executive strategic planning. Previously, she worked as the Manager of Corporate Strategy & Sustainability at a Canadian airport. There she developed and led a department that managed the Master Plan, five-year strategic plan, community infrastructure outreach, and the Corporate Sustainability Strategy. Her many international work experiences include research in global happiness, environmental assessments in Costa Rica, and municipal resource planning with a Canadian city. She has also worked as a data analyst with an international institution.

She was born and raised in the Middle East, completed her bachelors degree in Germany, and completed her research Masters in Resource Management in Canada. Since the completion of her Masters, Hana has been certified as an Environmental Professional and is a Certified Associate in Project Management.

She is the recent winner of Leading Ladies of Aerospace Advocate award and was elected to a North American aviation steering committee.

Follow Hana Galal on Instagram @grandtech2090 and keep watch for her upcoming book on the future of love and happiness.

Support the charity where she teaches English at https://eltoc.ca/donate.

CHAPTER 8

Creative Women in Science Have the Edge

Inga Leigh Gelford

"Always leave enough time in your life to do something that makes you happy, satisfied, even joyous. That has more of an effect on economic well-being than any other single factor." ~Paul Hawken

---

What would your life look like if you were able to be a leader in your field, attract and retain colleague collaborations, and have employees that adored you with a solid loyalty to you personally, as well as your projects?

What would it be worth to get to the core of how people tick and what's going on below the surface, thus enabling access to the power of really connecting with others? To develop the ability to influence people in ways you haven't before? What if you were able to help them to shine and stand out in their field? This is a superpower few scientists have. One of the ways this is accomplished is by actively accessing your creative side, even if you don't think you have one (everyone does). You may be surprised and amazed at what opens up when there is regular time devoted to creative expression. Logical, linear mind becomes balanced by intuitive, heart centered creative flow. Masculine balanced with the feminine.

Scientists are needed right now to bring a heart centered vision of solutions into the physical plane, and to also become powerful communicators. I'm on a mission to anchor the vision of what is possible and to disrupt the patterns of lack and limitation, by modeling flow state abundance and a devotion to creative solutions. I desire to inspire and move people to find a creative outlet of some form. I intend to stir your soul into creative action. The world is requesting we step up now. We are meant for more. Get ready to see who you are ready to become.

"Let's disrupt this world with possibilities." ~Lisa Nichols

One of the benefits of creating some form of art and being a phenomenal scientist are that it sparks intuitive connections. Creative expression as a process done with no end product in mind, or deemed necessary, is an exercise in wiring our brains to crossover to both hemispheres. We create for ourselves, to get in flow state, to relax and obtain mental fitness and flexibility. There often is a good dose of science and math in art and music. Creativity fosters connection, heart-based relationships, and opens collaborations that might not have been possible before. Creative expression draws you forth into becoming the person who can make a deep impact. When you are in alignment with your creative, intuitive self, you exude joy, and that is magnetic.

"Intuitive intelligence, I believe, will be the most powerful and sought after skill of the next decade-especially in the workplace. Intuitive intelligence seems to draw out insights, guidance and even a level of genius that some say come from beyond your brain." ~Vishen Lakiani

---

Creative, heart-based scientists make better leaders and have better retention and loyalty because they are more able to truly connect with others. Who you become, as you tap into your creative expression, is the key. This affects everyone around you, especially colleagues and employees, and extends to your personal relationships. This is an opportunity to embrace something bigger than yourself or your projects.

"Don't let the noise of others' opinions drown out your own inner voice. And most importantly, have the courage to follow your heart and intuition. They somehow already know what you truly want to become. Everything else is secondary." ~Steve Jobs

---

Jean Houston, an American author involved in the human potential movement, relays this story:

"I met Einstein when I was eight years old, when I went to PS6 where they took us to meet the great elders of the time. Helen Keller was one, Einstein was another. We were trotted across the river to Princeton where we sat down in a room with a big board filled with equations and this funny old man comes in, you know, he has a lot of hair, seems a little vague. He's got on a red sock and a blue sock. 'Yaah! What is your question?' So, there's a smart Alec case, one raises a hand and said, 'All right. Mr. Einstein, how can I get to be as smart as you?' He said, 'Read fairy tales!' We did not like that answer at all. So another smart Alec says, 'Mr. Einstein, how could we get to be smarter than you?'

'Ah! Read more fairy tales.' Well, I got a chance to go off to the side and talk to him for a few minutes. I said, 'Mr. Einstein, you're talking about the imagination, aren't you?' 'Yaah! You've got it, my child, it's imagination. Everybody thinks I'm so brilliant in mathematics. It's my students who do all that work, I'm a very poor mathematician. But I have a great imagination. I write with a beam of light, I go through the cosmos. I understand imagination. That's what it is!' Well, he was famous for saying that the imagination was more important than data, than what we think of as Intelligence."

Imagination and creativity are the foundation of all momentous breakthroughs. Einstein saw and embraced this. My experience also supports fostering imagination and being in your full creative bandwidth.

Working in the scientific field for decades, I've often noticed a particular work ethic that pervades the workplace. It is a sort of competition for those who work the hardest and longest, late nights gaining special brownie points. Bosses encourage this with an attitude that they own you because of a paycheck, so the work must get done no matter the personal cost. Maybe some people go along with this simply because they like testing their metal. For others, it seems they feel like they are never good enough, so this is their way of trying to prove they are. However, this usually excludes time for creative expression and for self-care, such as walks in nature. I also noticed professors often were not trained in people skills or management, so most were lacking in that regard. This sometimes resulted in very toxic work environments.

I remember having deep feelings of inadequacy and any feeling of success was all too fleeting. Feelings of accomplishment were all too quickly replaced with the thought that I would finally feel self-acceptance with the next big achievement. The cycle would go on and on with the never-ending thought of not quite ever being good enough, never quite there. One way I've gotten out of this self-defeating cycle and into feeling truly alive and full of self-worth is through daily creative expression. It becomes a current in my body, that insists on being expressed. When I was a child, I loved horses so much I pretended to be one, running around on all fours, whinnying and galloping around the house or yard. Of course I looked terribly silly and my siblings teased me mercilessly. Yet, I could not stop myself from doing it, even in the face of ridicule. I was flowing the life spark energy and power of what a horse represented to me. I find the creative spark to be the same, animating the whole of our life, and a habit well worth its weight in gold. It activates Life Force Energy and is our natural state of being. Each of us is a unique creative signature, a shimmering embodiment of extraordinary DNA. Solutions for the world's issues spring from tapping into the creative wellspring.

Below are six keys to establishing this habit:

1. Take a moment to identify your deep passion and to be introspective as to what truly lights you up. Breathe, move, spend time in nature or go for a long walk. Pause and listen, then capture what comes up and act on it.
2. Create a sacred space and time for creating visual art, writing stories, songs or articles, playing music, planting gardens, cooking, woodworking or any other form of creative expression. Invest in the supplies, spaces or workshops that allow you to implement your creative yearnings. If you are not sure, pick something anyway, you can change it later and you have started a momentum. Creating at the same time, place, and possibly having a music playlist, is ideal as it facilitates being in creator mode.
3. Surround yourself with people creating what you aspire to create, those who share your passions, be inspired and learn from them.
4. Some additional tools to flow creatively are: standing barefoot on the grass to ground your energy, use breath to stimulate the

mind and access the creative stream, adopt a power stance with conviction of purpose, create some bad art to get going, look at others for inspiration (go to art galleries, etc.), pretend you are a famous artist/musician/chef and already have those skills, thus adopting a new identity. Tap into a source of deep creativity that is limitless, be all in, decide, commit and have resolve of purpose.

5. Identify the saboteurs that may show up. The Judge: The Perfectionist/Stickler, Hyper Achiever, Avoider and the Hyper-rational. Embrace your Sage powers: Empathize, Explore, Innovate, Navigate and Activate when you find yourself tempted to forgo your time with yourself to create.
6. Make sure you are having fun! Activate your joy centers and everything benefits, including your scientific research. Dance, be silly, have epic adventures, notice beauty all around you, get excited by life! This also has the added benefit of increasing your immune system and overall wellbeing.

Place a daily time in your schedule now, create a time tracker or use an app to record your time spent creating. Remember, even one minute of time, just a few brush strokes on your painting or actions on your project, provides continuity and connection and gets the creation done even if slowly, and that's okay. It's the journey and who you are becoming that is the key more than a finished product. Remember, failing along the way is all part of the process and so is not taking yourself too seriously.

There is a saying that a good life is the enemy of a great life. I have a good life as a wine chemist. I am leaning into a great life as a change maker, visionary and leader of what it means to be creative. I am showing up as an entrepreneur, a writer, a visual artist and a transformational coach. I'm done playing small and remaining invisible. It's time to be seen and make a difference.

Inga Leigh Gelford

Inga Leigh Gelford has been immersed in creating visual art since she was a small child. Her education, however, has been in science, with a MS in Microbiology and currently works as a wine chemist. She is on a mission to be a facilitator of transformations. This involves inspiring people to identify as a creative, an artist, including, or maybe especially, women in science.

She sees creative expression as intrinsic to the purpose of why we are here. Views creativity, love, and joy as facets of the same gem, which is the life force energy that animates everything.

"Through the arts, we find humanity, connect with each other and learn empathy. The arts-singing, dancing, creating-are a basic human activity." ~Ronni Lacroute-Friend of the arts, co-founder of Willakenzie Estate winery

---

Inga is here to ignite potential, activate the DNA of creative flow and amplify brilliance. She desires to inspire and move people to find a creative outlet of some form. She intends to stir the soul into passionate creative action. From this activated creative space we find solutions, envision new possibilities, and enable the creation of a new paradigm of human existence that is heart based and has love and kindness as a contagion.

Connect with Inga at ingaccc@gmail.com.

CHAPTER 9

This Life Is Ours to Choose

Kristi Broom

Haven't always been this way
I wasn't born a renegade
I've felt alone, still feel afraid
I stumble through it anyway
~Pink*

I did not intend to have a career in technology. In my high school "what do you want to do with your life" dreams, I was an attorney, like the high powered, hit 'em where it hurts, expressing that zinger of a "he's guilty" line of questioning or closing statement, attorneys on TV. I was a fairly quiet, reserved, and most definitely not popular adolescent, which I imagine played into that desire. I chose my expensive private university because it was only one of three law schools in the state, and the only one with a 3-3 program, meaning I would graduate in six years instead of seven. And then I took my first pre-law class in my freshman year, and hated it. I learned that law was less about the perfect argument, that courtroom drama scenes are fairly uncommon, and that my job would likely entail hours of billable study of case law precedent. And so I pivoted. I liked my History class, and that became my new major. The challenge was that post-graduation I knew I didn't want to go to Grad School or teach, so that left me needing to find what I wanted to do with my life.

I was extremely fortunate to land a job with an amazing mentor. It was in the education industry, and she took me with her into corporate training at a new company. She was always one to let me stretch myself, both in my role and also into new roles within her purview. Though I started as an instructional designer, when I wanted to be a trainer, she let me try it. She gave me my first leadership role. Little did I know, those seeds planted all those years back would grow into where I am today. They blossomed into career journeys that involved teaching people how to succeed and to believe in themselves; to be their cheerleader, mentor and practical guide so they can navigate the complexities they experience. I am forever grateful to Shari for planting those seeds.

After she left, I had a series of good and bad bosses as most people do, and a mostly successful, still steep, career trajectory. After about a decade, the pivot that ultimately led me into technology showed up. I was running a team of eLearning Designers who had all the cool software along with a heavy workload that accompanied it. I needed a way to distribute the workload, and with some tutoring, was able to build a workflow tool and dashboard. A few months later, one of my colleagues – the one who connected me to those who helped me build those tools – was taking on a stretch assignment himself, and asked me to take over his technical team. I remember replying with a quick no, and "I can't even spell HTML." He convinced me that because of my natural curiosity about, and ability to play with technology, I was the perfect person, and my career trajectory changed. I did learn how to spell and use HTML, and built several other cool tools. Several months later, the same colleague was taking on another stretch role to build an innovation team, and asked me to lead the technology vertical. It was a resounding yes this time, and I have been there since.

Over the past decade, my career has continued to rise, though I did find new ladders which took me on some sidesteps to new paths. I continued to have some good bosses some bad bosses, roles that challenged me, and others where I found easy success. Those seeds planted decades ago carried through every role; no matter my own success, I strived to teach others to come with me. I am grateful to have had high performing teams in almost every case, maybe because I put great focus on building capability and confidence in those around me.

As I grew into senior and executive roles in tech, I started to realize that the world around me looked quite… male. I struggled a bit as each executive above me seemed to be another man whose qualifications looked a lot like mine, though the levels of admiration for them seemed much higher. I distinctly remember a moment when I was experiencing a lot of organizational change, and feeling like each new leader we brought in was not only a man, but a man from the outside. During that time, I had committed to supporting an event in which the opening keynote was Cheryl Reeve, the only coach in the Twin Cities with a championship team (of women), and yet we rarely hear about them. Cheryl said one thing that resonated: "when a man applies for a job, it's assumed he can do it; when women apply, they need to prove themselves."

That statement set off a spark in me, because Cheryl Reeves is IT in basketball, but she only gets credit for being IT in *women's* basketball. I didn't want to be that person. And yet as I look around, I am sometimes that person in technology; that lone woman representing, speaking for, and being a beacon for the aspiring women doing the hard work in organizations.

I reached out to a very senior executive woman and mentor to share what I was feeling. It didn't take me very long into our conversation before I lost it. Not angry, yelling, screaming lost it. I mean crying – ugly crying – lost it. And this amazing woman told me many things; among them: "if you ever need me to tell you you are a badass, call me because you are."

The truth is I need to be told that – either by myself, or by someone else. I have a necklace that hangs from my ring light that says You are a Badass. I have read the book of that title many times. Sometimes I have to think about it multiple times a day. I have playlists of songs that I play on repeat; Pink is among the Badass Women whose lyrics soothe my beleaguered, weary, imposter-syndrome-driven mind when I need it most.

I read recently that reframing the situation can help one to feel more powerful. In this case, being the only senior woman among men does make me feel a great sense of accomplishment, and power, and purpose. And then another spark came in a TV series about a woman of color who had to prove herself when the men were given the same opportunities without proof. It was a line delivered by another woman who said "maybe we need to support each other the way they do."

I wish someone would have told me that
this life is ours to choose
No one's handing you the keys, or a book
with all the rules
The little that I know I'll tell to you
~Pink*

In all of my years as a woman in the workplace, one of the things I've struggled with mightily is finding the words to say (especially when not given airspace, or when my volume is vastly drowned out by those louder than me). In all of the situations in which I struggle, the common theme is voice, or lack thereof, at being able to express my needs.

Don't get me wrong: there are replies readily available in my mind. Whether it is upbringing, or my reserved nature, or societal expectations for women in the workplace, I self-edit most of them to remain only in my inside-my-head voice.

And the truth is, I'm pretty good with words, especially in writing. But often in the moment, I don't say anything at all. Maybe it's because I'm not sure that in Technology, where 1s and 0s are literally the (code) language, that poetic, analogy-driven, stories are heard.

I've begun to think about what impact I can make, both for my daughter, who I hope doesn't experience gender bias, and for my sons, who often ask what they can do to help change the gender equity reality that we have lived in for a very long time.

My daughter's story is also one of pivots, and it is worth telling here. I remember when she was young telling her to "use her words." It's advice parents often give to get children away from temper tantrums and into more productive ways to express their feelings. As she grew into a teenager and eager babysitter of young kids, she said it often as well. Fast-forward to college, when we thought she'd be an early childhood educator teaching kids about words and lots of other skills, she found a love in Speech Language Pathology. Her graduation cap read, 'I teach kids to talk back.' Like those of Cheryl Reeves, those words stuck with me, and likely for a purpose.

I've been writing my "career advice for my younger self" book in my head for quite a while, capturing the stories I've experienced. Through this process, I uncovered analysis of the words I should have used, would have used, if I could find my voice. I want to teach others to talk back.

I've decided that a bigger story needs to be told; one I can certainly contribute to given my career experiences and the advice I can pay forward as a result of them. But also one that is more practical and actionable, for people of all genders who are seeking to eliminate gender inequalities in the workplace. My goal is to give women the words they need to talk back. I'm talking actual words, such as "excuse me, I was talking" when interrupted. Or, "how is what you said different from what I just said?" when someone feels the need to restate a thought or idea. I also want to give men the words as well, so they can not only be allies but active advocates, using their voices to say the same words. Most importantly, I want to contribute to a larger community of women who support one another. Because that movie script writer was right; we do need to support each other as women, in order to rise up to what we deserve.

I have so many more stories to tell, and as I talk to women in senior positions and those starting out, I know there are so many more stories out there. If you are interested in telling yours, I am ready to listen. If you are willing to share your story, or your practical words to guide young women, I'm here.

My goal is to teach women to stand up to the inequities and everyday situations that currently prevent us from achieving true equity in the workplace. I want to give real words to use in the situations when they may not come easily. I want to be the cheerleader on the sidelines of every woman who learns to use these words, so that together we can make meaningful change for women in Technology and all other disciplines. I want to teach women to talk back. I hope you will join me.

*Lyrics from "All I Know so Far" by Pink

Kristi Broom

Kristi Broom has spent 25+ years in learning and talent, rising to increased levels of accountability in each role. It was because of an aptitude for exploring technology solutions, as well as the ability to be both deeply strategic and forward thinking, and deeply planful in executing solutions to the finish line, that she was asked to lead a team of technologists, and then drive innovation for learning technology, and she hasn't looked back since.

Kristi is a leader and visionary, a coach and a cheerleader, thriving in her own success but most happy when those around her succeed. In recent years, Kristi has taken on leadership roles in DEIB, focusing on bringing in perspectives from global cultures, and supporting the next generation of aspiring women leaders.

In addition to work, Kristi is a certified yoga instructor, speaker, writer, avid reader, and scrapbooker of the memories shared with her husband, three children, and fur babies, including her current puppy, Dakota.

This book is a first foray into writing words to empower and support women to rise up in their careers. A spark has been lit within Kristi to write and do more, and she would love to hear your stories, perhaps as content for another book focused on practical words that women can use when faced with difficult workplace interactions.

Connect with Kristi on LinkedIn at www.linkedin.com/kristibroom.

CHAPTER 10

Queen Victoria Syndrome: the Veterinary Edition

Legend Thurman

"Dependable... constant... capable of weathering any storm. Elements based on foundations of logic and diplomacy... some driven by faith but always passion...."

This particular mindset I have developed over the past nine years as I have progressed through two scientific degrees with the end goal of becoming a veterinarian. Now that I am a few months shy of gaining this humble, yet empowering, award classification, it is nice to reflect back on the journey that was traveled to get here: one that was filled with winding roads and unexpected but rewarding decisions.

My education began at the Catholic University of America in Washington, DC, where I studied biology, chemistry, and theology/ religious studies; this combination would not only fulfill the requirements for admittance into veterinary school, but also challenged myself to appreciate and leave with a well-rounded, liberal arts education. Additionally, having a strong faith propelled me to not only better understand the unique relationship between science and religion, but apply it daily. As I referenced in *The Book I Read* (my second collaborative anthology), the initiatives that surround both science and faith are inter-connected, striving for us to question and study how they both support one another in balance. For without faith, we all live in a world of doubt, but without science, we have nothing to question.

CUA's environment was immersed with the typical scientific set-up with standard lectures, seminars, and laboratory sessions surrounding subjects like genetics, microbiology, organic chemistry, physics, and a two-part graduating thesis along with a very large Basilica sitting outside on the front lawn. The days were long with several occasions of wanting to toss the material out the window combined with classic imposter syndrome but also very fond memories. My junior year, the year I applied to over twenty veterinary schools even with the credit card yelling at me, I vividly remember looking through the plethora of documents at my desk and seeing the application for the Royal Veterinary College, the number one veterinary school in the world, and thinking: "this is never going to happen, but let's just put it in the pile." Fast forward six months later where I am driving in downtown DC with one of my best friends, Vincent Fung, and learned that I gained admittance to this institution making it an option. A place that was not only highly ranked but several hundred miles across an ocean was now on the table along with some other really good schools including The Ohio State University. Safe to say, though, the suitcases were packed and a one-way ticket to England was purchased.

Let's now change scenery to a completely new atmosphere. Imagine you are walking through the most prestigious academic institution for your subject field for graduate school and around every corner is an opportunity to not only meet someone new but also a challenge. Entering the UK educational system was certainly not a walk in the park as I had to adopt new methodologies for both learning and application of practical skills, but as I near the conclusion of this chapter, I can confidently say that it has made me a better veterinarian constantly aware of the facts, communication, my empathy levels, and boundaries. While my undergrad prepared me with the scientific, foundational knowledge and the freedom to cross-examine, my graduate degree taught me that your skills are multi-faceted and easily transferable. For example, when the average person hears the occupation, "veterinarian" and are questioned with what their skills and positions entail, most answer with something in the clinical realm. However, what if I played devil's advocate and introduced politics into the equation?

An average veterinary course runs anywhere from four to seven years depending on where you obtain your education and involves ample time split between pre-clinical and clinical entities. It is during these periods where one progresses through the foundations of anatomy and physiology while slowly escalating to clinical reasoning, problem-solving, and rotating between different departments in veterinary hospitals, farms, and more. After graduation from a program, a large percentage of individuals choose to go down the path of clinical practice by entering straight into the workforce or completing an internship that transitions into a residency training program in order to obtain board certification in a particular subject. For example: one can become board certified in veterinary cardiology, oncology, etc.; it is the same process and subject areas that most human physicians undertake within their professional realm. However, the uncommon theme within veterinary medicine or in any of the clinical professions is the unnoticed aspect of transferable skills.

During my second year at the Royal Veterinary College, I discovered my passion for veterinary government work as I ended up chairing the Government Affairs Committee for the National Student American Veterinary Medical Association. This avenue entailed learning about legislative initiatives, public policy, and advocacy all while promoting a one-health agenda, a scheme that recognizes the connection between humans, animals, and the environment. While these terms seem broad, they encompass a myriad of platforms within the field such as agriculture, animal welfare, transportation efforts, drug regulations, food standards, and more. Additionally, besides my keen interest in all species and the enthusiasm to constantly learn more effective ways to better the profession, there has always been an affection on my end for organized leadership. What better way to combine my education, collaboration efforts with several other occupations, and core passions at the end of the day!

Over the next three years, my time was evenly split between clinical and non-clinical work. I have circulated through the standard small animal, equine, and farm rotations mixed with external placements at several institutions both in America and the United Kingdom. Some of these included the House of Lords in Parliament; the Veterinary Policy Research Foundation; the Department for Environment, Food, and Rural Affairs; the Veterinary Medicines Directorate; the National Pork Producers Council; the American Veterinary Medical Association; the British Veterinary Association; and the Royal College of Veterinary Surgeons, just to name a few. The days spent in the governmental field were always individualistic and included work on the spectrum of veterinary medicine from ethical debates to advancement of the profession to public health risks, affecting aspects like communication, trade, growth, and more. New skills that I also have acquired along the way and in my toolbox include:

1. Writing Memorandums

2. Preparing and Giving Briefing to Senior Political Officials
3. Advising political members, small stakeholders, industry, heads of departments of regulators
4. Writing White Papers and One Papers
5. Attending meetings with various protocols, goals, and mindsets
6. Truly learning the impact a veterinarian can make on the laws that are enacted and the decisions made on a local, state, federal, and international level

After participating in these activities and seeing the overall impact they had on our profession, I knew that I had found my niche especially after my briefing I wrote in support of the Glue Traps Bill in Parliament on one of my placements was used for discussion and helped lead to its passage. It even gave me the confidence to write my veterinary thesis on the Animal Sentience Act here in the UK, completing a legislative analysis.

Moving forward, the interesting fact is that anytime I engage with others about my plans when I qualify, I am given the opportunity to educate individuals on the many places you can find a veterinarian working in society, not just in clinics, on farms, or in zoos. My favorite response is when I hear, "so you don't actually want to be a real vet and work with animals." A veterinary professional does not just simply possess the clinical knowledge needed to treat your animals or perform surgeries, but also holds advanced communication skills, entrepreneurial endeavors, emotional empathy, and a drive to make a difference. One of my favorite hobbies happens to be networking at conferences where I am able to connect with a plethora of individuals each with their own story to tell and dreams to aspire to. Myself and my colleagues in the STEM/STEAM field each make up a piece of a bigger puzzle working to advance the direction this profession is headed, and we all have our place. We currently live in a learned society where one has to continue to grow with a profession that is evolving, but the direction it is heading is up to the community of individuals in its diverse makeup.

In terms of where I am headed for employment, the next few months should uncover the doors that will be opening to the next chapter soon; however, as a sneak peak, I am focusing my time and efforts on aiming for jobs in the DMV (DC, Maryland, Virginia) or Chicago areas either in industry, federal/state government systems, or corporate organizations carrying out political, organized leadership, and organized medicine initiatives. And if you are wondering about the clinical aspect of my career, I can promise you that it will not be diminished in any capacity as locuming and relief shifts in the local community, on a mixed scale, are also within my prerogative and interest. The possibilities are endless with always the option to start over and begin something anew.

The veterinary profession, like so many others, carries its own hardships that easily make people want to leave as quickly as they enter the field. However, I am not here to conclude on that somber note, but rather to remind you that your skills are transferable and passions are contagious. There is no one correct course to go down regarding employment or happiness, but rather the freedom and tenacity of will to find your place in a world seeking contentment. Society may push an idealistic paradigm of the definition of a "true occupation," but the modalities surrounding every personal journey stems from the steering of one's own course in life. Queen Victoria, one of my favorite leaders, who ruled the English Monarchy for over 63 years even during times of turmoil and societal views of abdication said it best: *"The important thing is not what they think of me, but what I think of them."*

Build your own platform and dare to lead with grit and grace aiming to be a person of value, not simply success in an environment where you are meant to thrive and not just survive. My name is Dr. Legend Thurman, a governmental veterinarian; what's your story?

Legend Thurman

2X International #1 Bestselling Author and a native of Washington, Pennsylvania, and Washington, DC, Legend Thurman is currently a Doctor of Veterinary Medicine Candidate at the Royal Veterinary College aspiring to be a governmental veterinarian advocating to give a voice to those who cannot speak for themselves while being rooted in servant leadership.

Featured with the AVMA, SAVMA, DVM360, Vets for Success, Vet Candy, Women Action Takers, and several others, some of her previous work has included animal/veterinary legislative initiatives, public policy, and advocacy on a local, national, and international level.

Legend is also a firm believer in overcoming some of the struggles humanity faces in terms of acceptance, self-image, vulnerability, and imposter syndrome among veterinary professionals and society as a whole.

When not working on her degree, you can find her at the Basilica of the National Shrine of the Immaculate Conception in DC, traveling, or spending quality time with those she loves. She is set to qualify in July 2023 and plans to move back to the USA to carry out her mission of creating everlasting change for all creatures great and small in alignment with the AVMA Agenda.

Connect with Legend at https://www.linkedin.com/in/legend-thurman.

CHAPTER 11

Making Cybersecurity More Diverse, Equitable and Inclusive

Maggie Calle

Introduction

Little did I know back in 1994 what my life in Canada would be. I arrived in Canada with a hundred dollars in my pocket and my little brother in tow. From there, I was on social assistance/welfare for a year, then student loans and part-time jobs to get me through college, then my very first job in Technology, a few Cybersecurity executive roles in large financials, and now I'm a Chief Information Security Officer (CISO) of a very reputable organization with an amazing culture. For all intents and purposes, I've made it and I am privileged to represent minorities in Cybersecurity. Although the number of women in Cybersecurity and in top leadership positions such as a CISO role are improving, we can do better. An ISC2 study found that nearly 24% of the cybersecurity workforce are women while men continue to outnumber women and pay disparity still exists. Also, women in cyber maintain higher levels of education and are making their way to top leadership roles in higher numbers than before. (ISC2, 2018).[1] Additionally, the pandemic reversed this progress when working mothers, women in senior leadership roles, and black women opted out of the workforce. (McKinsey, 2021).[2]

But have I really made it? Sometimes I question myself as it has been a challenging road that has given me some perspective on possible solutions to help others navigate to top leadership roles. Cybersecurity is interesting and never gets boring; I feel a huge responsibility to inspire the next generation, especially women who look like me.

As you might imagine, I have some good, bad, weird, and funny stories that helped me become a better leader. At a young age, I came to realize that my team lead did not want any competition in our male-dominated environment and she had crowned herself the "Queen Bee," making my life impossible and even suggesting to my manager that although I was competent, intelligent, and high performing, she suspected my hyper-focus may be the result of using drugs – her clue was my fidgeting feet and my ability to complete my deliverables really fast. As my manager shared this feedback from my team lead, I knew then I wasn't going to succeed in this environment and I managed to get a job outside the department, of which my manager was very supportive.

After this experience, I transferred over to the Information Security department in a management role working in Network Security, Business Continuity, Crisis Management, and Operations Resiliency. My new manager was really great, supportive of my continuing education towards a degree, and very experienced in operations. Sometimes he showed genuine concern as I was involved with high profile projects, taking up to three courses per semester at night school, working a lot of extra hours at nighttime implementing solutions, all while I was planning my wedding. At my first performance review, my manager said he was given excellent 360 feedback on me from peers, managers, and stakeholders. He thought one was interesting and shared: "Maggie has tons of energy to multitask; the most accurate way to describe her is that she is like an Energizer bunny that keeps on going." I got a bit concerned thinking, uh oh, here we go again. Now I think this is actually a very accurate and classic way to describe an individual with ADHD, but in a positive way. What I mean is that I didn't feel that my manager's feedback was downplaying my performance, contributions or work. I was simply a very hyper-focused individual who fidgeted a lot because that was a way to channel my energy, which is actually quite typical of an ADHD person since physical activity increases the level of neurotransmitters dopamine and norepinephrine similar to ADHD medications. I was finally at a place where I was being recognized for my work; a place where I could be myself without repercussions and flourish.

Cybersecurity Challenges

How a Hustle Culture is a major detractor for Diversity, Equity, and Inclusion Efforts

Although the purpose of this chapter is to inspire women to get into cybersecurity, we can't ignore the systemic challenges in this field. For a long time, cybersecurity was glamorized as a field for very technical people, hackers in black hoodies whose lifestyle is to just focus on hacking and nothing else. Going to conferences acting as tough dudes, partying and drinking hard were key aspects of being in cybersecurity. This was not the type of environment that would attract a soccer mom or the type of professionals that would get the Board's attention to get budgets approved. I don't mean any disrespect to hackers and the amazing work they do, but I am simply highlighting a challenge we have created in the cybersecurity industry by glamorizing a hustle culture while relegating diversity, equity, and inclusion efforts. Let's be honest, would a man or woman with children be attracted to an industry if they were told they had to hustle 24/7 adopting a lifestyle where work is their priority over their family, children, or anything else? Childless men and women who have other interests outside work would not be attracted to a field that consumes their livelihood.

In the past two decades, I've seen this culture shifting where security professionals attend cybersecurity conferences that include technical, strategic, and beginner tracks. Some events are geared towards senior executives to hone their skills to talk to business stakeholders and to address the Board of Directors. The cybersecurity culture is shifting in a positive direction where collaboration and strong communication skills are extremely important and there are a variety of roles for technical and non-technical talent. I have been part of contributing to this cultural shift where my security teams are co-creating value with business stakeholders and no longer lurking in the shadows; this is a huge step towards changing the mentality that only hackers or technical people can work in cybersecurity.

When I discussed this glamorization of the hustle culture with a leader in my organization, his response was that "we have chosen this lifestyle because we are passionate about cyber and we knew what we were getting into," which left me disappointed. Leaders need to resolve our workforce burnout. If we want to run 24/7 security operations, we should be outsourcing operations or hiring three shifts versus expecting the same employees to work 24/7. Leaders can change the cybersecurity culture to be more welcoming of a diverse workforce inclusive of those with children and without, neurodiverse and neurotypical.

Do we actually have a cybersecurity talent gap?

For three years in a row, the ISC2 published their report on the Cyber Workforce study highlighting a gap of 4 million cyber professionals in

2019, 3.12 million gap in 2020, and 2.72 million gap in 2021. (ISC2, 2021)[3]. Although the talent gap is closing, we still have a lot of security professionals also having extreme difficulty finding jobs in cybersecurity, not to mention the turnover and poor retention that we have in this field.

Once, I joined an organization where I had to skip my March break family vacation in my first three months of employment in order to fulfill a couple of commitments (out of many) that were overdue for two years. Leaders should not misrepresent jobs and have unrealistic expectations from workers (men or women) and definitely should not overcommit their teams just so they can get their trophies and promotions; a company with this reputation will have issues finding talent. When I sent my husband and children off on the vacation, I made the decision to never prioritize work over my family and I submitted my resignation shortly thereafter.

We also have poorly written job descriptions requiring a laundry list of responsibilities and cybersecurity certifications for junior roles. Additionally, I have personally experienced situations where I had to get HR to move on the hiring process very quickly as to not lose knowledgeable candidates who are in high demand. I often ended up with candidates taking an offer elsewhere because our HR process was too slow.

Do we actually have a talent gap or an inability to attract the next generation of cybersecurity professionals? How do we create a diverse talent pipeline when our Millennials and Gen Z have already decided cybersecurity is not their hustle and they would rather be YouTubers, video gamers, podcasters, Amazon sellers and all other cool professions. I think the answer is creating an inclusive work environment and great culture that attract a diverse workforce.

Does Networking through Social Media work?

I am going to start this section with a resounding YES and I have evidence to prove it. For a past job search, I decided to leverage LinkedIn extensively. First, I created the alerts for my desired target jobs, I then applied for jobs and contacted a total of 42 LinkedIn contracts from which an impressive 66% responded. I contacted either: 1) the job poster; 2) the hiring manager who I found by researching on LinkedIn; or 3) LinkedIn connections who worked in the organization who I asked to pass my resume to the hiring manager.

This was a very successful search resulting in a total of three job offers from amazing organizations. The details follow: A) nine out of thirteen leaders I have indirectly worked with in the industry responded with interest, resulting in a job offer that I accepted, B) ten out of fourteen leaders I had worked with directly in the industry responded with follow-up interviews; and C) nine out of fifteen contacts who I have never met before responded positively and two resulted in two job offers.

Simple and Effective Solutions

Cybersecurity leaders should actively leverage social networks, partner with post-secondary institutions, and establish rotational programs to bring entry level resources into the industry. I have personally found and hired many brilliant and amazing candidates straight out of LinkedIn, universities through the Magnet Network, or by attending college and university Career Day Events.

A few simple things we should all do: 1) let's improve job descriptions and get rid of the hustle culture; 2) work with HR to do better pre-screening of candidates; and 3) don't focus on red flags but potential. Keep in mind that cybersecurity workers move jobs very frequently to find better work/life balance, to learn new skills, and not just for money. In fact, I once took a lower rank job because it was ten minutes from home when my kids were younger. I also hired fifteen new graduates from diverse university and college programs to rotate them into different functions for short periods as a training mechanism with amazing results. Where HR sees red flags, I see well-rounded individuals in charge of their careers who've picked up great cyber skills by moving around and prioritizing properly.

Executive Search Firms must provide better services to women trying to make it into cybersecurity executive ranks. In my experience, I have seen these firms act as a barrier for women impacting Diversity, Equity, and Inclusion (DEI) efforts. These firms should review their historical stats and percentage of women they've placed in executive cybersecurity positions as compared to men. With numbers in hand, they can implement training to better screen candidates with DEI in mind.

Conclusion

I know a lot of leaders and peers that are proactively making a difference and solving the systemic issues discussed in this chapter; together, we can continue to effect positive changes in the cybersecurity industry. The biggest lesson I learned in my journey is to not stay in places where I am not appreciated or wanted and that I should take charge of my career.

In closing, let's create self-awareness amongst cybersecurity leaders to solve the issues we face. The next generation of cybersecurity professionals are watching us as we represent what they could become, and we don't want to deter them from coming into this wonderful profession. Finally, a big thanks to the women that have paved the way and shattered the glass ceiling before me; I would not be successful today if it wasn't for their trailblazing efforts.

Maggie Calle

Maggie Calle currently serves as the CISO at Varicent. For over two decades, she has held security leadership roles in the financial, insurance, and retail sectors as well as Technology companies. She has successfully established a "tone at the top" in the management and oversight of cybersecurity and risk management programs supporting business objectives, innovation, and digital transformations.

She has been recognized as Women to Watch by Risky Women organization, Canada's Top 20 Women in Cybersecurity by ITWC, Top Influencer in Cybersecurity by IFSEC Global, and Cybersecurity Woman of the Year by SiberX. Regularly speaking at cybersecurity conferences, she also mentors and promotes STEM education in her community and as a guest speaker at many educational institutions.

She holds active CISSP and PMP certifications, a Bachelor of Commerce Information Technology Management degree from Toronto Metropolitan University, an MBA in Risk Management and Corporate Governance from Athabasca University, a Masters Certificate in Project Management from Schulich University, a Diploma in Computer Programming and Analysis from Seneca College, and a Human Resources Graduate Certificate from Seneca College.

Connect with Maggie at https://www.linkedin.com/in/magalicalle.

CHAPTER 12

I Am Woman! I Can Do Anything!

Melodie Donovan

If you read my story in my first book, *Invisible No More; Invincible Forever More*, you know part of my story of breaking down barriers and the challenges along the way. I will recap and fill in some gaps and tell you my experience of becoming a woman in the technology industry.

I grew up in a small, rural town in Indiana. Boys were boys; girls were girls. As a female, I wasn't supposed to be interested in graphic design, shop class, drafting, chemistry or physics. Being a small-town school, there weren't' many resources to provide a big variety of classes. In 1982, my senior year, our high school received new Commodore 64 desktop computers! Yep, Commodore computers! They decided to offer two sections of classes each day. Each class period would allow five students to learn how to use and write programs on the new computers. To gain access to one of the classes, seniors were required to submit a written request to the principal in order to be allowed into the class. The students had to state their plans to attend college and study computer science to be considered for one of the ten open spots.

I was lucky to have been one of the students chosen for one of the classes. There were two boys and three girls in our section. It was an independent class that was monitored by the principal himself. Google didn't exist then. Some basic information was provided so we were on our own to collaborate with each other and to somehow write a computer program. We all took up learning Dungeons and Dragons and writing code in DOS. This sparked my curiosity about computers and programming. I entered college declaring a major in Computer Science. It certainly wasn't an easy road. I realized in my first quarter in college that Computer Science included a lot of math classes. Math was not my strong suit. During this same time, my oldest sister started dating her future husband. He was finishing his master's degree in a program called Management Information Systems. I was intrigued and started asking questions and investigating. I would summarize Management Information Systems as the business application of computers. Professionals in this field would work with companies to determine their computer needs and match their needs with the ever-changing computer equipment and applications that would solve and improve their business processes. It piqued my interest, so I switched my major to Management Information Systems and kept marching forward. I enjoyed the program and looked forward to graduation. In college, I never felt any preferential treatment being given to guys. We all seemed to be on an even playing field… at least at that time.

After graduation, I felt I had a lot to prove to my dad and to myself. He didn't think girls needed to go to college. In fact, he expected me to drop out once I found a boyfriend and get my degree in "Mrs." There were a lot of girls in that "program" at college, but I wasn't one of them. Sure, I wanted a husband and family, but I wanted to have a successful career, just as much, if not more. Interviews for computer jobs were few and far between. I just needed to get a job and earn a living. I moved to Indianapolis and ended up starting post-graduate work in retail. My long-convoluted path to a technology job began! From retail, I moved to an office management job. That job included maintaining the office computers and server for our accounting team and back office. Over the next fifteen years, I would have a few different jobs as office manager, controller, assistant buyer, and human resources administrator. All of those jobs were multifaceted and most contained duties to maintain and troubleshoot the company's computer systems or some part of technology. Throughout those years, I applied to technology jobs, but just didn't have any hands-on experience managing larger computer systems.

In 2010, another job change would be the biggest catalyst for embracing the technology industry and a dedicated position as a woman in technology. I started a new job as the Business Administrator. My main tasks were in the accounting realm of purchasing, travel and a little desktop support. An opportunity came to increase the technology work by assisting the Operations and Systems Director. It started with managing the users and permissions to our CRM and SharePoint applications. Inch by inch, I learned more about the computer applications being used. I did a lot of independent study by watching tutorials (thank you YouTube) and asked a lot of how-to questions. The Operations and Systems Director eventually left. While management took their time to fill the position, I stepped in to manage the applications in the interim. I would like to say that it was all smooth sailing from then on, but that was not the case. There was a male executive who, to this day, I have no idea why, didn't like me and THAT was an understatement! It was known throughout the company that he had a "hit" list and I found out I was on it. Not only did I have to keep working hard to prove myself, but I also had to watch my back. He wanted me out. It made me more determined than ever to dig my heels in and stand my ground. As a woman in technology, I felt very alone most of the time. You couldn't trust anyone. Most women will say the same. Not only were you battling men in the industry, but you also had to deal with other women in the industry. We were all after the same thing- that coveted ONE spot that only ONE of us was going to be able to obtain. Instead of allies, we were very much each other's biggest competition. Only in the last three to five years has there been a change in that thinking. Now, you will see women locking arms and saying let's get one of us to the top and she will then reach her hand down to help lift other women to that next level as well.

Don't get me wrong, all men are not bad, I was fortunate to have two managers that ran "defense" for me. They were the ones that saw my worth and experience. They shielded me as much as possible from the executive. They even managed to get me a raise! I actually felt I made some progress. It still wasn't enough to have them promote me. The two managers that protected me soon announced they were moving on to new companies. While the one executive who had it in for me also left the company, I thought "things were looking up." My boss, the IT Director, was leaving and I threw my name into the ring to be his replacement. That didn't happen. The new executive was personally connected to the old executive. He, too, wasn't going to allow me to advance in my role.

Your confidence takes a huge hit when you are consistently held down, not recognized for your work, and your boss even steals your credit. It took coaching and a little therapy to realize I deserved better than what my current employer was allowing me to experience. I had no respect for the executive staff nor could I trust them with my future career development. I found it very hard to interview and sell myself to perspective employers. I was constantly bombarded with negative self-talk and frustrated in a hostile work environment.

It's very important to surround yourself with colleagues who will lift you up, encourage you, and help you improve. I believe for every negative thought that is put in front of you and lands in your mind, it takes more than one-hundred positive encouraging thoughts to wipe out the negative thoughts. I don't want to give these people or any of my other naysayers any more time in this story. Suffice it to say, they put obstacles in my way and blocked me from advancing. I managed to overcome them all. They became challenges for me to learn and grow from. Now, they serve to add interesting details to my story. My giving any more attention to them here, could be considered man-bashing and making excuses.

My new job is a breath of fresh air. Although I am, most often, one woman among eight or ten men, I am respected for my experience and skills. There is still more work to do to make for an equal playing field for women in the technology industry. As women we need to continue to build our networking and collaboration efforts. If you are fortunate to make it to the C suite, look around and see where you can make a place for another woman in technology on your team. United we stand ~ Divided we fall.

Melodie Donovan

Melodie Donovan is an Award-Winning #1 International Bestselling Author, an IT Professional of twenty years, a bartender for six, and a serial entrepreneur for 11. Melodie launched her financial coaching business to help women with their journey to financial freedom. She is a Certified Financial Coach through Ramsey Financial Coaching. She became a travel agent and Traveling Vineyards wine guide. She enjoys learning more about wine, meeting other people, and traveling. Melodie is thankful for everyone who has encouraged her over the years. She is especially thankful for her children, Zach, and Casey. By encouraging her children so that they can be anything they want to be with a little hard work, a lot of faith and belief in themselves she has also learned to be kinder to herself and an encourager to herself and others. She is thankful to have another opportunity to contribute to another Action Takers Publishing book.

Connect with Melodie at https://MelodieInc.com.

CHAPTER 13

Chasing Storms

Mindy Maggio

I finally felt that the storm lifted. I've had glimpses of the blue sky many times, but never really soaked in the sun's rays until one day, in 2013, when I *handed over my heart* on the way home. That's a bit of a metaphor for what actually happened on a drive home to Phoenix from San Diego through the desert.

With twenty-five years as an IT Infrastructure-focused professional, I experienced a lot of weather. (*IT infrastructure is "the system of hardware, software, facilities and service components that support the delivery of business systems and IT-enabled processes"* according to Gartner.) As a woman in a male-dominated *segment* of a male-dominated *industry*, most of the time I honestly felt like one of the guys. I didn't think there was anything wrong with that, until I learned I was shutting off a part of me that held my superpowers.

In IT, storms were common. Some I knew were coming and could plan for; others that were never on the radar popped up. This is how I spent my career both in project management and managing operations.

I thrived during storms; I saw blue skies here and there, and felt the freedom when storms lifted over and over.

This is my story of when I started creating my own weather.

Yes, I was a storm chaser. Every project and team had its own weather. In my experience, there were always two ways to manage weather. I tried to lead with my superpowers and values of love and kindness rooted in connection, relatedness, curiosity, and possibility. Some, on the other hand, tried to lead with chaos, confusion, bullying, and deception. This is the kind of storm I found myself in quite frequently, which created a lot of *internal* storms for me. I butted up against this behavior time and time again, and worked to conform to it, because I thought that was what we were supposed to do. I often found myself soaked to the bone, "*standing on the corner, waiting for a light to come on*" (*Kings of Leon – Cold Desert*).

The lesson? The unique gifts that made me *me* were not something to be ignored. But instead, I pushed them down to *adjust* to what I saw around me. I did that my whole life. From the beginning, when I heard, "you throw like a girl," I was taught to throw like a boy. I changed. I adjusted. I pushed down what was natural. I was made to feel weak for the way I did something naturally. No wonder I struggled with my feelings, always felt off, that something wasn't right, and worked harder, proved myself, burned myself out.

But it worked! I was successful. I achieved many things. I moved up the ladder and was on my way to becoming a CIO (Chief Information Officer). Midway through my career, I was even honored with the *Needle with the Red Dot* award from Perot Systems. I was considered "*one in a million*" from my program director at the time.

"...and I tell them that we aren't just looking for a needle in a haystack. We are looking for needles with red dots."

~H. Ross Perot

Here's the other thing I had going for me. *Or did I?* I am of small stature, 5 feet 1 inches tall to be exact and, fortunately, looked very young as I aged. I was always looked upon and treated like a little sister (which in fact I am, of two older brothers, who toughened me up), or as a little girl; at least I often felt that way.

Because of this youthful gift of mine, I was overlooked several times for named leadership roles. Even though I had a track record of successful implementations and leadership qualities, was extremely personable, efficient, and known for my ability to bring order to chaos, that same program director told me, "*I couldn't put you above some of the older guys who have been around a long time. They wouldn't respond well to a young girl.*" I was 31. So, I carried on. I worked my tail off, unconsciously aware that I was in constant *proving-myself* mode.

Another example of being passed over happened a few years later in my career. My colleague and I wrote a proposal that outlined a large reorganization of our department. It was introducing centralized engineering and massive efficiencies in field services operations. We recommended ourselves to manage the restructuring and ongoing operations. It was well received, and we were super excited. The leadership decided to bring someone in from corporate to execute our plan. You could say that I once again felt like a little girl that never gets a break. What was I told was the reason? That I wouldn't be able to fire people, as part of the required cost reductions. I guess I was just too nice (that loving kindness that was my core value got in the way).

I was gathering amazing experience on these projects, both as an innovative and strategic leader. But I was not on the books as the official "leader" on many projects. It certainly didn't help me build my confidence or my resume.

One of the most impactful quotes that supports my experience is from Sheryl Sandberg that "men are promoted based on their potential while women are promoted based on their past accomplishments." As a woman in a leadership role, I made it my mission to see the potential in everyone, not just men, and give people a chance to try, to shine, to make mistakes, and to learn.

You can see in these very brief examples how I felt a lot of storms coming from inside me. These were the storms I sought shelter from but still got wet, worn out, or wounded. The storms inside included feeling small, not good enough, misunderstood, and underrepresented.

While there are external factors such as cultural belief systems that work against us succeeding, our challenges are what Claire Zammit, Ph.D., founder of Feminine Power, refers to as the "*inner glass ceiling.*" What I learned about my own inner glass ceiling is that I wasn't overlooked necessarily because I looked like a young girl; I was overlooked because I *felt* like a little girl. Until I started to *feel* confident, valuable, worthy, and mature, others would view me this way. I'm not saying that would've been easy, but if I'd had a mentor or support system that helped me see these blind spots, I may have shown up differently. Once you are aware, only then can you start to make shifts that shatter your inner glass ceiling.

Where do you begin, you might ask? It starts with understanding you can create your own experience (aka *create your own weather)* and make new choices.

I started to actively participate in my life on that day in 2013, on that drive home from San Diego. I will never forget the moment. My husband and I were driving home from San Diego. It was during a long stretch of desert highway, staring out the window, when the song *Cold Desert* (Kings of Leon) came on the radio and this incredible sense of peace came over me. I felt it so deeply that it felt like it was coming from my soul. The peace was a complete feeling of freedom. It was the feeling of a storm lifting and blue skies peeking through. I looked over at my husband and said, "I want to feel this way every moment of every day. This is what life should feel like." It's about feeling love, wholeness, and connected to something greater.

As I sat with that feeling, I pondered thoughts that questioned why I was doing what I was doing – spending most of my life at work, no longer feeling satisfied with the status quo. Although I was successful, it came with stress, self-doubt, overwhelm and a lot of proving myself, over and over. Although I made good money, held a good position, I wasn't living the life that brought me peace and ease every moment of every day. In that moment, I made my first choice toward peace. I emailed a friend who was a life coach.

Working with her, I learned to be intentional in the experience I wanted to have in this one life. I was guided to surround myself with a support system of people who provided different perspectives and experiences and challenged me to stretch out of my comfort zone. I also worked with a spiritual coach who helped me create and connect with a deeper awareness of myself and an understanding of my ego.

I worked with both coaches (and several since) to connect with what I really wanted and what I was no longer willing to tolerate. I began the process of defining what my personal freedoms were.

I like to think of these freedoms in the following ways. *Freedom from…* (beliefs, thoughts, societal constraints we believed to be true but aren't), *Freedom to…* (do what we want to do), and *Freedom to be…* (who we're meant to be).

"Not all storms come to disrupt your life; some come to clear your path." ~Paulo Coelho

---

What if every storm you are going through is preparing you for the next one? When I began to look at my experiences weathering storms, I learned proper gear is essential to staying warm and dry… or *centered and thriving*. I now help others find the balance they are seeking and the wisdom in their storms.

I hope the following mindset shifts and principles from my *Personal Freedom Toolkit* are helpful for others to lead, weather storms, and begin to create their own weather.

Be intentional. Living intentionally is thinking about life with a deliberate and purposeful mindset. It's about being conscious of the choices and decisions you are making when it comes to your thoughts and feelings. You get to choose what you think and how you feel.

Create a deeper connection with yourself. Get to know yourself by making the time and creating the space to connect with your deeper desires, and understand what your values, priorities, and personal freedoms are.

Design systems that support your growth. This includes support for your emotional, mental, and spiritual growth. Meditate, journal, do yoga, dance, listen to music, get a mentor, a coach, join a women's circle. Seek out those who will help you *safely* see your blind spots, support your courage, and help you get out of your comfort zone.

Allow your curiosity to stretch and expand your knowledge. There is never a dumb question, so don't be afraid to ask.

Authentic confidence is fueled by self-love. Your authentic self is your *best* self and staying true to who you are is courageous.

Embrace the balance of masculine and feminine qualities that make you uniquely you. Don't push down those parts of you that you were taught were weak. Allow them to become the power behind what others can tangibly see in you. As an example, I can be sweet, kind, and loving *AND* smart, strategic, and strong.

Understand how you are participating in how others see you. What limiting beliefs might you be radiating that might be holding you back?

Feel connected to something greater. We are all connected and part of this vast life-positive Universe. When we set intentions, we are in co-creation with something greater. Think of yourself like a beautiful flower in a grand garden. Your only job is to grow and to blossom.

"*The world needs all its flowers, every one.*" ~Jon Kabat Zinn

---

I believe we can create our own weather by intentionally living a well-balanced, purpose-fueled, vision-driven life. Then we can start to play our bigger game and uplevel our impact and influence. This is a game-changer for women who lead and have the desire to make an impact.

Celebrating you and your best life! You can always find me "*Where the Skies Are Blue*" (The Lumineers).

Mindy Maggio

Mindy Maggio is a coach, mentor and transformational facilitator, working with women leaders on mindset and empowerment.

Her background is in Information Technology, with over 25 years of experience in technical project management and infrastructure operations. She started her own company as an IT consultant in 2014. In 2020, she made a shift where she connected her calling with her career. She founded the Bridge-Ville Networking & Coaching Community, a space for women leaders to connect, collaborate, and experience community.

She works with women leaders who want balance, empowering them to actively participate in their life, how they lead, and leave their legacy through impact.

She is passionate about creating connection and community for women leaders that:

- supports growth and confidence
- challenges conventional sources of power
- inspires creating balance and understanding personal freedoms

She believes it is our responsibility as leaders and as women to create the change we want to see in the world. We can lead in the direction of where we want the world to go by modeling purposeful leadership styles that transform people, teams, and organizations.

One of the ways she does this is with a monthly ROAR, which is a 'Round of Applause Ritual.' This is a free live event where women leaders gather to voice their brilliance through their monthly wins. They focus on the strengths that create these wins and stand for one another in support and amplification.

Connect with Mindy at www.bridge-ville.com.

CHAPTER 14

Moving Toward Success, One Bit at a Time

Nicole Scheffler

As a woman in tech, I have enjoyed the last twenty years building complex networking and security architectures for some of the largest tech companies. At the dawn of my next chapter, I am redefining success on my own terms and invite you to join me. I love sharing proven success principles and techniques as part of my mission to spark success for women in tech by leading and serving. My hope is that these stories inspire you.

Program your GPS and fulfill your potential

Much like a Global Positioning System (GPS) helps people navigate turn by turn to their destination, there are systems and people to help guide each turn on your route to success through much of your life. To graduate from high school or college, counselors help you create the course plan needed to graduate. They incorporate your passions into these plans so you explore new things to learn and grow. This 'mentorship' continues throughout life in various forms from credit card pay off plans to wedding plans. In many situations, there is someone riding along helping you stay accountable.

Then, there comes a time when no one helps you shape a success plan and sometimes no one is riding along to guide you. Without owning your destiny, it's like sitting in a car hoping to get somewhere great without anything programmed into your GPS. I have had setbacks where I have had to recalculate my route or even change my destination altogether. After you reach a certain point in life, it becomes up to you to program your own GPS. You own your success.

In 2005, I was selected as the valedictorian student speaker for my graduating class. It was a huge stretch for me to do this and I remember feeling so sick on stage, overwhelmed while looking into the crowd before I gave a speech on this exact lesson on life accountability.

Ultimately, it's up to you to program your GPS based on your passion and goals. Don't get left sitting in your car with nowhere to go. Life is full of adventures.

Throughout my career, I have been on some amazing journeys from speaking on over forty stages, hosting over one hundred podcast episodes, and even sitting in the front row for a Sting concert in Hawaii. They all started with a destination. I encourage you to program your GPS today and keep dreaming of your next big adventure.

What is your next destination to program in YOUR GPS?

Remix your future and embrace confidence

I have always loved music! I spent my college years at University of North Texas, where music was often at the center of the scene. I went to a weekly DJ party and came to realize there were very few women DJs. I decided to grab a couple turntables, dig up a solid vinyl collection, and step up to the tables.

This was my first real dance with technology, since I provided all my own gear and ran my own production. I had to set up and network all the equipment, troubleshoot sound issues, and ultimately create an experience for the dance floor through technology. It was also my first time to lean into such a male dominated field.

You hear the phrase "step out of your comfort zone," but sometimes that just means it's time to take your own stage. From this experience, I grew confident taking on challenges with technology as one of the few women on deck. This skill ended up serving me well so far in my twenty-year tech career. Living out of your comfort zone may help you recognize skills and passions you didn't know you had, but confidence is something that can last forever.

How can you step out of your comfort zone?

Bit by bit to build your empire

It's easy in a 'compare culture,' heightened by social media and filters, to feel lacking at times. You see other women rising in their careers, the ones making big moves, or those who have already arrived at a destination you can only imagine. That is why it is key to remember that the only person you should compare yourself to is the person you were yesterday.

It's the small habits you do every day that can help you prepare for your future. Success is truly at the intersection of preparedness and opportunity. You can only take one step forward each day, bit by bit. What you practice in private, shows in public. No success was built in a day, so don't compare yourself to others.

When I write code, it is a super agile process. I make one change, compile the code, and review my result. I rarely get it right the first time, but regardless, I always take a step forward. Each time I present my product at work, I get new questions and a variety of perspectives as feedback. As a result, my presentation gets that much better the next time around. Every event I attend, I add a new woman to my network. These relationships add up and come in handy when I need it most. By the time people see you in the spotlight, you will have been building to that excellence bit by bit over time.

Embrace that process in your own life. One bit at a time, you can work toward your dreams. This is done with intention, like setting your GPS, and with relentless pursuit. These bits are paving the path of your unique journey during your time here. Make them count. Bit by bit, you will build your legacy.

Reflect on your own accomplishments and cherish every bit.

By sharing these stories with you, I hope you feel inspired to pursue your dreams in any STEM field. My own success would not be possible without all the tech divas before me that paved the way. I can only hope this insight inspires you in the same way I have been inspired by all of them.

It takes all of us to encourage and push each other beyond the hard days and to move towards the bright future waiting for us all. Together we can do more.

Nicole Scheffler

Nicole Scheffler has been a Tech Diva for about 20 years professionally. By day, she works in Cyber Security as a Solutions Director at Palo Alto Networks. At night, she is dedicated to spark success for women in technology careers with the Tech Diva Success Collection.

She started her career in programming at a startup, then spent about 15 years of her career at Cisco Systems in engineering and in Strategy and Planning. Nicole is the author of three Best Selling books prior to this one, *Pillars of Success with Jack Canfield, 1 Habit of the Greatest Leaders*, and *1 Habit to Thrive in a Post Covid World.*

Nicole co-founded the Diva Tech Talk podcast in 2015, which has now won 8 Clarion awards, providing a library of women sharing their inspirational and diverse career journeys to spur more women to enter and/or stay in technology. In addition, Nicole founded the Tech Diva Success podcast designed to offer more diverse perspectives that all help ignite a real excitement for a career in technology. She speaks often on success and mindset as a certified Success coach and offers goal setting courses held with the Tech Diva Success Club.

She would love for you to check out the agile body of work created to empower tech divas: a woman that believes in herself and is ready to take action to make a big impact in a male dominated field. Nicole loves exploring future technologies and spending time outdoors with her family.

Nicole dedicates this to her own tech divas in training, Norah and Stella.

Connect with Nicole at https://www.techdivasuccess.com.

CHAPTER 15

All Is Not Lost

Niki Hall

A single career, like my father's working with one company his entire life, is no longer realistic. It's not as fulfilling as one would think. It becomes less meaningful to one's intellect over time and the stimulus you once felt turns into habitual discipline.

It's become normal for a person to have pursued a multiple career path over the span of their working life. For my age group, to live through three different careers over a working lifespan is normal. This statistic will change depending on your age category.

People naturally have more than one aptitude they could choose from to develop into a vocation. We are diverse beings and so are our natural abilities to do and create. At an early age, I knew about three prevalent prominent areas of interest within me. An engineering desire and my loving people aptitude were obvious and out front in my character as a toddler and young child. The writing came a little later, during my high school years. I still write and my passion for people has continually developed into my presently being a Consultant and

Facilitator in the field of Mindset Coaching, for those who are looking for advancement in business and wealth. But I am not here to share these two professions.

What I want to share is an area where the workforce has changed over the years. That is with the presence of women in fields that once were solely populated by men. I have always wanted to know how things worked. I fantasized about having a mind like great thinkers that were creative and innovative. I liked being a problem-solver, had a skill for seeing detail, and I was a good listener. All these things, plus a few more characteristics, led me to think 'engineer.' So, I decided I wanted to be an engineer when I grew up. And that desire never left me.

Now I want to share with you my journey of longing to be an engineer, and where I landed over 20 plus years later. This is my story.

I don't believe I was walking yet, but I was crawling and I could sit up nicely. My hair was long and dark. I was in a very pretty dress, tights and patent leather shoes. My parents were determined to take a portrait of their "beautiful little baby girl." Unfortunately, they couldn't get that shot of me because I wouldn't stop crying. Over time, they figured out I didn't want the fancy doll they were trying to sway my attention with. You see, my father's electronic equipment was holding down the front corners of the blanket that I was sitting on and that is what I wanted. My parents eventually figured it out. The result was, every time they showed that portrait, they had to explain why I was so happy chewing on knobs that obviously came off electrical equipment.

Clearly that made a statement about me. I always gravitated toward machinery instead of dolls and sweet or delicate toys. And yet my parents continued to put me in these beautiful dresses, maintained my long pretty hair, and kept me in coloured tights and patent leather shoes.

We lived in a great neighborhood, diversified in culture, with a lot of kids. One morning, my parents gave me permission to go out and play with the other kids, but told me to be careful of my nice pretty dress, tights and shoes. I quickly agreed to that and hightailed it outside to play with my friends as fast as I could.

I could see my dad kept peeking out the back window to discreetly keep an eye on me. At one point he saw me in a dirty wheelbarrow, crammed in with three other kids, while another kid wheeled us around the backyard. The next time he looked out, I was pushing the three kids around. Then, the next time my dad looked out, to his surprise, two of the kids I was playing with and 2 of the bigger kids and I were in the middle of an all-out fist fight and I was holding my own, right in the middle of it all.

At that point, my parents accepted the thought that I wasn't going to be their beautiful little girl that they could showcase. I was a kid that liked to play outside and fight with the boys. That is when they cut my hair short and let me wear my brother's hand-me-down clothes to play with the kids in the neighborhood. I was becoming the person I was meant to be.

Now let's jump to my sixth to eighth grade years. My parents would occasionally catch me at various points of disassembling or re-assembling all types of machinery, anything I could get my hands on. I wanted to understand how it worked. It was when they caught me with both the washing machine and the dryer taken apart at the same time that my parents relented one more time to accept that I had an aptitude that they were not going to be able to stop or divert with other interests. I'm moving closer to that engineering identity.

During summer after the eighth grade, I had a lot of conversations with my parents. You see, at that time, girls going into high school would go into typing, shorthand, and such things to become an efficient secretary and the boys would go into the tech wing for woodworking, electronics, automotive, and drafting classes, for various industrial jobs or trades. My parents finally relented to let me go talk to the school principal to let me into the tech wing. The result there was that I was the first girl in the tech wing. I was now on the right track to live the life and get the career that best suited me. My parents finally accepted that all I wanted to do was go to school and get a Ph.D. in engineering. I was very focused and driven.

This was working out for several years, until someone literally walked across my path. I knew that young man was meant for me. In a microsecond my entire life plan turned upside down. Nineteen months later, I was married. The next year, I was pregnant. The next year, I was a mother. My life was completely different than I had foreseen and planned.

Now, I'm at the tail end of the baby boomer generation. At the extreme other end is a sub-sect we referred to as the old boys' network. They put a lot of burden on others that were looking to become accomplished and looking to move ahead in various fields of occupation. For myself, the old boys' network would close ranks and ran a closed system of relationships that made opportunity for those who weren't within that small and calculated community difficult for acknowledgement and certain advancement or for creating their own opportunity. This caused challenge, confusion, and hardship. I was among those people. I went through a lot of challenges and holding my own to keep moving in the direction I saw fit for myself and my family.

Also, I didn't finish my education plan. After my child was a certain age, I tried to go back to school to pick up where I left off, but I was told I couldn't because too much time had lapsed. I would have to start all over again, which I wasn't prepared to do. I felt all was lost in my effort to become an engineer. Still, I had the engineering aptitude with an ability to write. I was still at a stage in my life where I was holding the balance of my own interests, as well as maintaining my duties as a good wife, being an attentive mother, and now wanted to contribute to the family income. I had to find a job that could use my skills. I found myself in jobs where I began report or business writing for industrial companies. I still kept a technical toe in the water, as it were, because I still possessed that enthusiasm toward engineering. It was obvious how interested and how much I wanted to be a part of engineering; I still had a natural ability for details and precision, add that to my critical thinking, creativity, and innovativeness. I could also write and was an active listener. This landed me work with salesmen who were selling machinery and such to big companies.

I became a professional business report writer. I did that during my married life. Then one morning my husband was bringing me my morning coffee with a sweet morning kiss, then by 2 o'clock that afternoon my marriage was over. I likened it to being a crystal figurine of a woman standing, imagine there is a little area, that if you hit it just right, the figurine could smash into a million pieces. Which I did. That is how I felt at 2 o'clock that afternoon. As quickly as I walked that man into my life, twenty-eight years later, I walked him just as quickly out of my life. Again, my life had completely and radically changed.

I needed to find a new career. I needed to be completely financially self-sufficient and I lived in a big city. The cost of living was high. I somehow found myself among business owners and engineers. They were still able to see the aptitude in me: my technical passion, interest, and aptitude. You see, I thought because I wasn't able to finish my schooling that I couldn't call on anything I learned or my natural ability because I didn't have the degree. I didn't have that piece of paper stating my acumen. Nevertheless, this very successful businessman called three of his best engineers into his office with us and asked them if they applied any of the things they learnt in school on the job. All three of them said not really, for various reasons. This gave me confidence because what they were telling me is it's who you are. So, I took my place among the good old boys' network, to be their technical writer. How is that for a turn of events? I worked among a group of five to seven men for thirteen years. I worked with some amazingly intelligent, a few I would call absolutely genius. They trusted me with talking technical information and felt confident that I could understand and deliver the information accurately. Which I did.

In the long run. I was still able to use my technical ability to see and understand the technology of design, of buildings and new build projects. Honestly, I was among some international geniuses. I was able to breathe in, live, and be part of such a diverse set of projects, that if I became an engineer my discipline would have caused me to be part of a less diverse group of projects. Through becoming a technical writer, my experience was vast and I was very privileged to be part of some amazing new inventive projects. My sense of being an engineer was accomplished beyond the measure I could have obtained through my own means. I recently left that world and I finally feel complete and accomplished there.

The point for me is I know many of us who go toward an engineering or STEM career don't follow through because of other things that took priority. You can go back, even if you don't have a diploma, that sacred piece of paper. The job you do get in that direction may please and fulfill you more than if you went the conventional way. Don't feel less than and don't fear for lack of knowing. You got this. All was not lost because you had to deviate from your goal for a period of time. Go for it! All is not lost!

Niki Hall

Niki Hall is a Mindset Coach.

Realizing her personal ability to help people, she opened a self-help school. She later wrote a book entitled, *Building Up - Thoughts Expressed During the Readjustment of Self*, a book on Change and Self-Actualization, where her first printing sold out within days of its release. This catapulted her into public speaking and workshop engagements.

Now Niki has hung a shingle out to help people break through their limited beliefs and to achieve more prosperity in their health, wealth, or business.

Connect with Niki at https://www.facebook.com/niki.hall.148.

CHAPTER 16

Moving Mountains

Sharleen Gatcha

As a proud Canadian, I am fortunate to have the Rocky Mountains in my own backyard. Regrettably, I did not come to fully appreciate this until many decades into in my life; however, I am grateful that I now realize the benefits of spending time amidst them, appreciating the sights and sounds, and enjoying the beautiful scenery. I share this as it provides some context for one of my favorite quotes that I want to begin my story.

"I stand on the sacrifices of a million women before me thinking, what can I do to make this mountain taller so the women after me can see farther." ~Canadian poet Rupi Kaur

---

I have spent my 30-year career in the energy sector, primarily in the electricity industry where women represent only one quarter of the workforce. During this timeframe, I have witnessed immense change as the energy industry has evolved from the use of coal-based fuels to renewables, and now to move away from historically carbon intensive practices to net-zero emissions to meet the targets set by what is commonly referred to as "The Paris Agreement" to address the world's desire to accelerate climate action, sparking a global energy transition that will challenge everyone to think and act differently to protect the world we live in. Unfortunately, despite this global crisis, what hasn't changed are the barriers that women continue to face in the workplace that result from and in the under-representation of women in STEM careers.

I have experienced firsthand many of these barriers and have had to make many sacrifices that I know are common to many women in practically every workplace. Despite an evidently high degree of skill and ambition, women in STEM careers have additional challenges given their reduced numbers in the overall workforce, resulting in a lack of role models, mentors, and opportunities to apply their knowledge. Some of the barriers and challenges I have experienced include:

- Gender bias: widespread and in all forms including being considered "aggressive" for being openly ambitious, speaking up in meetings and speaking out about inequities.
- The "Motherhood Penalty": being passed over for promotions because I was a working mom and 'couldn't be relied upon.' Because I was not permitted to work a flexible schedule, I was forced to reduce my role to part-time although my workload was not reduced, so I completed my work in four days while I was only paid a part-time salary and performance pay was also pro-rated despite the fact I completed the same amount of work as a full-time employee (in four days vs. five days) and was able to maintain high performance ratings.
- The "Broken Rung": struggling to get on that first rung of the corporate ladder (first management position) primarily because of the barriers previously mentioned.

- The "Double Bind": In addition to the Motherhood Penalty mentioned previously, I was often told I wasn't ready for promotion but was not provided any tangible or actionable feedback to be promoted. In essence, my reward for good work was being given more work! I often received conflicting messages that my performance was exceptional; however, it seemed that no matter what I did, there was always another threshold to meet, or another hurdle put in my way. Surprisingly, my male colleagues did not have these same challenges.
- Sexual harassment: subtle and sometimes blatant, from leaders and other co-workers as well as clients I worked with. I left one organization specifically because of this and expressed this in an exit interview. The organization did nothing about it and my male leader continued this behaviour until his retirement many years later, affecting the careers of many women after me.
- Judgment and discrimination from other women: when I finally did obtain a well-earned promotion, I learned that my *female* colleagues were spreading rumours that I only received it because I had an inappropriate relationship with the male leader that promoted me.
- Emotional burden: A persistent and constant feeling of overwhelm and exhaustion for the emotional labour involved with managing all of the above challenges and feeling as though I had to fight for every opportunity every step of the way....

I share my experiences with the reader to illustrate the misogynistic environment that is so entrenched in our culture and pervasive for women. This is to demonstrate why there is good reason women are under-represented in STEM careers. It is time for women to claim their space and speak out about the challenges they face so they can succeed in the careers they want. It is my conviction that women should not have to fight for the right to be treated equal, to choose to raise a family, and to be treated with dignity and respect, especially in the workplace!

In Canada, it was only ninety-three years ago that women won the right to be considered "persons" in what is referred to historically as the 'Person's Case.' The case was started by a group of women activists, now known as the Famous Five (Nellie McClung, Henrietta Muir Edwards, Irene Parlby, Louise McKinney, and Emily Murphy), who objected to a Supreme Court of Canada ruling that women were not "persons" and, as such, were not allowed to serve in the Senate. The Famous Five challenged the law in 1929 and women were legally recognized as "persons." In Canada, we celebrate this victory and the resolve of the Famous Five each year on October 18.

There is no doubt the battle for women to become 'persons' was hard fought and won; however, the battle for equal rights for women in the present day continues to rage on. Despite tremendous strides, we are still far from gender equality. Although 82% of women aged twenty-five to fifty-four now participate in Canada's workforce, they are still underrepresented in leadership roles. Women hold only 25% of Vice President positions, and 15% of CEO positions. Women tend to work in industries that reflect traditional gender roles, specifically healthcare and social assistance, educational services, hospitality, and food services. Even in female dominated industries, women tend to occupy lower-level jobs. In the food-services industry, for example, nearly 60% of chefs are men while 71% of servers are women. According to Statistics Canada, as of 2021, the gender pay gap is 0.89, which means women make 89 cents of every dollar men make. The UN's Human Rights Committee has raised concerns about "persisting inequalities between women and men" in Canada, including the "high level of the pay gap" and its disproportionate effect on low-income women, racialized women, and Indigenous women.

According to the World Economic Forum's 2022 Global Gender Gap Report[1], it will take 132 years to reach full parity at the current rate of progress! With the risks associated with the global climate crisis, it has been said that we are more likely to destroy the planet within this timeframe than we are to reach gender parity. Astounding.

I shared the Rupi Kaur quote at the outset and my experience in the workplace because I want the sacrifices I have made and the challenges I have overcome in my career to empower women. By sharing my challenges and discussing openly the barriers and challenges that many women face in the workplace, I hope that I can help pave the way for the women that follow to make their journey to fulfill their career aspirations shorter, less challenging, or more rewarding. If I can do this, then my effort will have been worth it.

In 2019, I started a non-profit organization to support women working in Alberta's power industry. Women+Power has grown to over 700 members and has hosted 27 events and programs with over 3500 participants.

Women+Power is about supporting women. Our board includes high-profile women and men from the energy industry who have experienced many of the barriers and challenges I mentioned. We know firsthand that when women are provided opportunities to come together in a supportive environment to share their challenges and successes, when they are empowered to excel in their careers and are recognized and celebrated for their efforts and achievements, transformational change is the result.

I found that by sharing my challenges with other women in the industry it has helped me to understand I wasn't alone and allowed me to develop strong connections that helped me to learn and grow, to build confidence and ultimately advance my career. I wanted to create this opportunity for other women and that is what led me to create

1 https://www.weforum.org/reports/global-gender-gap-report-2022/in-full

Women+Power. My vulnerability in this regard has opened doors for other women to do the same and many women have told me touching stories about how grateful they are for the community of support we have developed so they don't feel alone and have a network they can reach out to for opportunities to connect, learn, support, inspire, and recognize each other to achieve their full potential both professionally and personally.

Women+Power's tag line is: "Instead of a corporate ladder that women feel they must climb alone, Women+Power is a community that will climb mountains together."

Together, we can make the mountains taller and climb them together. And, as in the Rupi Kaur quote, we can see farther and create opportunities that are limitless.

There is much work to do to get there.

In November 2021, McKinsey & Co. issued the "Gender Diversity at Work in Canada" report[2]. The report discusses the findings from research related to the challenges and inequalities I have discussed, which have been amplified by the COVID-19 pandemic, such as:

- Throughout the pandemic, there has been an increase in household responsibilities. While many partnerships split the additional work equally, a higher proportion of women than men reported taking on all or most of the additional work.
- More women report being exhausted, or chronically stressed.
- Women reported concerns about the repercussions of requesting or taking advantage of workplace flexibility, despite benefitting from options such as remote work.
- Women reported providing more pandemic-related support to team members than ever before.
- Women also stepped up outside of pandemic-related support, championing DE&I progress within their organizations.

2 https://www.mckinsey.com/ca/overview/gender-diversity-at-work-in-canada

In terms of making progress towards DE&I, this requires action across five dimensions: 1) understanding, aspirations and commitment from management; 2) governance and accountability; 3) people policies, processes, and systems; 4) supportive workplace models and programs; and 5) inclusive mindsets and behaviours.

Investing in diversity, equity, and inclusion creates company-wide benefits, from overall performance to a stronger sense of belonging for all. Companies can address these challenges and take advantage of the shift that will occur in processes and policies. To do so, they must meaningfully invest in actions that demonstrate progress toward their DE&I goals. This will demand an inclusion-centred work environment, shifts in all employees' mindsets, and a clear understanding of how DE&I benefits everyone. Companies must hold themselves accountable for their commitments and implement supportive programs and policies that ensure all employees have equitable access to workplace opportunities — now and in the future.

This is even more important for women in STEM. As mentioned earlier, a global energy transition is underway with growing consensus regarding the need to move away from historically carbon intensive practices to net-zero emissions. This transition offers a unique and dynamic opportunity for policymakers and organizations to be proactive and leverage the diverse mindsets and human skills that women can provide in addressing and managing challenges that many organizations are experiencing post-COVID, and to achieve gender equity and pay parity that is long overdue.

To achieve this, the energy sector must move towards becoming an inclusive, innovative space for diverse thinking and leadership that benefits not only women, but everyone. Industry leaders, provincial and federal governments, and educational institutions have the opportunity to be proactive regarding gender equity as we transition our energy systems to prepare for a carbon neutral future.

Alberta's economy, and that of the rest of the world, is undergoing a period of immense change and opportunity. Historically and currently, women have faced and continue to face a variety of barriers to equitable participation in the energy sector. These inequities will persist through the energy transition if they are not actively addressed today by leaders across the industry, in government and education.

Research has proven that diversity in the workforce:

- Fosters communication and collaboration
- Breeds innovation
- Improves decision making
- Enhances performance and increases profits

To keep pace with the evolving demands of the energy sector, and more specifically to realize the benefits that women bring to STEM careers, companies must become consciously inclusive and encourage diversity in the workplace.

This is only possible if we ensure our workplaces do not tolerate gender bias and discrimination, by strengthening the processes for handling these issues, and emphasizing the need for male-allyship in the workplace by providing skills and training opportunities for men to acquire tools to participate in shaping gender-inclusive work environments. This includes setting meaningful targets for gender and other kinds of diversity at all levels of the organization, especially in leadership and/or decision-making positions.

My contribution as Founder & CEO of Women+Power over the past few years has reignited my passion for DE&I so much so that I have decided to pursue a career in this area, to work with companies in the energy sector to create inclusive, innovative spaces for diverse thinking and thought leadership that benefits not only women, but everyone to ensure that we are able to meet the targets set out in The

Paris Agreement and address the global climate crisis that is upon us. Women are an important part of the solution, and it is my hope that this is the opportunity we have all been waiting for, to reach that summit and stand together in unison to celebrate our accomplishments and all of the effort it took to get there.

We can move mountains together.

Sharleen Gatcha

As a corporate strategist intent on continuous improvement with more than 25 years of experience in the energy sector, Sharleen Gatcha has an established career that has allowed her to witness the full scope of the energy transition commencing in the early 1990s. She has extensive knowledge in many areas in the sector as demonstrated by her professional experience which includes a broad range of corporate leadership roles with key organizations. In late 2021, she started her own consulting business to offer her energy sector expertise and passion for inclusive diversity to various clients and she is now looking to broaden her reach to further develop her career.

As founder and CEO of Women+Power, she is also committed to empowering women and promoting gender diversity in the power industry in Alberta. You can often find her participating on panels or in projects that encourage discussions that get to the heart of issues that women face in their careers and her passion has led to the development of initiatives, programs, and resources that women need to succeed in their careers.

Sharleen has her ICD.D and has extensive corporate governance experience, having worked with numerous corporate boards throughout her career. In addition, Sharleen's volunteer board experience spans two decades and includes work with several diverse not-for-profit organizations. She is currently President of Lean In Calgary and Co-Chair of the Governance Committee for Light Up the World.

She has taken numerous leadership courses throughout her career and is currently participating in the Bold & Visible program, attaining her Brain Story Certification offered by the Alberta Family Wellness Initiative, and completing the Indigenous Canada Certificate Program through the University of Alberta.

Connect with Sharleen at https://www.linkedin.com/in/sharleen-gatcha.

CHAPTER 17

Transforming the Leadership status Quo in STEM: Embracing the Journey and Taking the Lead

Dr Susan McGinty

"I believe female leadership can create powerful, positive and far-reaching impact. We can keep talking about the lack of women in STEM leadership and the impact it is having on STEM capability, or we can start doing something about it now."

~Dr Susan McGinty

Growing up, I was curious about the why behind everything. I was drawn to science. It provided knowledge behind life, the universe, and how humans and the world function, which I found fascinating. I loved health science, anything to do with the human body. I poured over my mum's nursing and drug reference books for hours, captivated by the knowledge they held and the utility of that knowledge. I studied chemistry and advanced maths in the upper levels of high school; I loved organic chemistry and had a high academic record.

I wanted to apply knowledge that fascinated me to help others. I was encouraged into STEM tertiary education by my chemistry teacher, who introduced me to a new degree in medicinal chemistry (drug design), the first of its kind in Australia. This really excited me. I studied all forms of chemistry, biology, anatomy, and pharmacology, and how to design new molecules to interact at the cellular level to treat disease. I loved that I was learning how to create new scientific knowledge to positively impact people.

I enjoyed my first taste of research so much that I undertook a PhD. Through research, science offered exploration of curiosity, possibility, and the generation of new knowledge. For me, the power of new knowledge was its application to improve outcomes for society – impact. Upon reflection, I realise my desire to create long-lasting impact and change began manifesting in my early twenties, though I had only an implicit sense of this at the time.

After ten years in chemistry research, I moved to public service and the Defence realm. This allowed me to realise domestic and international impact through my scientific knowledge. It also introduced me to the concept of leadership and placed me on the path that led to my vision, and an even bigger contribution to STEM.

Through my experience in public service, I came to understand and develop a passion for leadership. Twenty-five years' experience in the world of STEM – through study, research, application to Defence policy and capability, and broader industry – has given me a deep understanding of the barriers holding back women in STEM and the impacts on STEM outcomes of the lack of gender diversity in STEM and STEM leadership.

I believe passionately in the strengths women bring to leadership and the power of diversity in leadership to tackle the STEM challenges we face as a society. My vision is equal representation of women at all leadership levels in STEM professions. To achieve this, I now work to disrupt and change the status quo for women as leaders in STEM by uplifting and empowering more women into leadership roles. I create impact by contributing my experience and expertise to develop enduring leadership in STEM women, so they can create impact in the world.

CHALLENGES

I didn't always have this vision; it was a product of my experience. While I enjoyed science research and envisioned it as my lifetime career, it was a difficult place for me. I was trying to establish a career based upon the linear academic career path that had been modelled to me: gain a PhD, undertake postdoctoral research overseas, publish research, gain grant funding, establish myself as a researcher, then obtain a lecturing position. That wasn't how my story went, despite receiving an award for my PhD.

In my science education and research career, I had very limited access to female role models, mentors, support networks, or career guidance. I didn't realise how important these were for providing access to opportunities, teaching me about barriers for women in STEM, showing me how to navigate the system, shaping my ideas of success, and understanding different career options. I saw men gaining career support from more senior men and thought I just had to work harder. I had a strategic mind, but this wasn't appreciated in the research space. Of the few women within my visibility, most were putting their careers on hold to have a family or putting having a family on hold to establish a career. I knew I would have to make that choice and chose family early in my career. What I didn't learn until many years later was the importance of female role models and networks in supporting women to navigate that discrimination.

Although I loved science research, I didn't see a career pathway that aligned with my choice of family. When I finished my PhD, I was already married, my husband and I already had our first home, and I had my husband's career to consider. I chose jobs at our local university that wouldn't require me to move away from home for weeks at a time or travel long distances daily, which I knew would limit my career options. I persevered with the scientific grant funding processes; however, I only made it to 'finalist.'

I was in the same situation that many STEM women find themselves in, suffering the cultural and structural gender biases present in STEM industries. But it wasn't until many years later that I understood the depth of the problem. When I became pregnant with my first child, into my third post-doctoral role and on a limited term contract, I suffered similar discrimination as many other women in science. I had to fight to get my contract renewed, and when my maternity leave finished, I wasn't permitted to return to my research role in a part-time capacity. And I didn't want to return full time, nor did I want someone telling me I had to. Other employment options in my location were very limited, so I undertook casual teaching and tutoring roles at the university. A week before my second child was born, I was interviewed for a full-time science manager role. I was offered the role the day after my daughter was born. A condition of accepting the role was that I commence within a few weeks. I chose not to take the job given the terms.

I had learnt over the course of that time that a science research career path wasn't a good fit for me because I didn't adhere to the expected norms of the academic research pathway. More importantly, I think I subconsciously realised that via research I wouldn't be guaranteed to achieve the impact I knew I wanted to achieve in the world. So, I sought alternate ways to apply my passion for science and moved to a role in the Department of Defence where I could apply my science knowledge and skills to national security. In the process, I learnt that there is more than one path to success, that my idea of success could adapt and grow with me, and that the pain of my early career was instrumental in the journey to what is now my bigger vision.

Leadership is now the centre of what I do and how I create impact in the world. However, my own leadership journey was challenging. As a scientist, leadership was not role modelled. I grew up thinking that leaders made decisions and told people what to do. It wasn't until after I found myself in leadership roles that I began to understand leadership as the complex process that it is and found some good leader role models.

Formal leadership development in science was non-existent. In Defence, as a civilian, I found limited access to formal leadership development training. I learnt as I made mistakes. I floundered. I had to teach myself and it took too long. I had some good mentors to guide me; however I didn't understand how important mentors were so I didn't seek them out, as I now advise others to. My organisation was male-dominated and there was a large military workforce. Leadership styles were predominantly masculine leadership styles. Male colleagues had role models, though not necessarily good ones. I noticed them being sponsored and networking – and getting the best opportunities because of it. Sitting at the intersection of being a woman in STEM and in Defence, there were almost no role models for me, or for the STEM women around me who lacked confidence and belief they had a meaningful career path. It lit a fire in me to know that the system didn't support STEM women, like me, to succeed and didn't teach them how to share their expertise, have their voice heard, and influence effectively. That fire smoldered for some time before it intensified.

TRIUMPHS

I was determined to be a good leader and learnt to be one. My leadership development occurred slowly, driven by my learning about how leadership works most effectively, improving my communication and influence, and developing other people to their highest potential. By 2015, I was mentoring and coaching the women around me, and giving them opportunities to grow their leadership skills. In 2016, I undertook a leadership training program that unlocked my ability to lead others in a way that aligned with me.

I am a firm believer in taking the lead when you are passionate about the need for change. I wanted to see more women in STEM leadership roles. In 2018, I decided I would create the leadership development for STEM women that was so desperately needed. I was determined to help others learn from my experiences, mistakes and learning, so 1) they could fast-track their own leadership development, 2) more STEM women could be uplifted and empowered into leadership roles at all levels, and 3) together we could start changing the status quo for women as leaders, and these female leaders could create significant impact on the world. My vision was created. My passion and purpose were strong. The fire was blazing.

I developed and piloted my first women's leadership program in 2019. It was research-informed and research-based and represented all the foundations of great leadership that I knew and continued to learn. It was more impactful than I anticipated. Participants of the pilot program were a mix of STEM and non-STEM professionals. I was surprised at the significant change it created in women personally and professionally, and many pilot participants were immediately promoted or given expanded leadership opportunities. This success reinforced my purpose and passion. I continued to deliver the program during 2020-22, continually adapting it, and have expanded my services to additional leadership programs for leaders at all levels, accessible webinar learning sessions, and executive leadership coaching. Over 2019-2022, I supported more than 350 women to develop their leadership confidence and capabilities through coaching and training.

In 2020, I created Aya Leadership as a vehicle to my vision of equal representation of women at all leadership levels in STEM professions. The word Aya comes from the African Adinkra fern, a symbol of endurance and resourcefulness, qualities I seek to instill in female leaders. I have undertaken a Master of Leadership and stepped into Aya Leadership full time in July 2022. Every day I live my mission to educate and support women to develop enduring leadership capability, to uplift and empower more women into leadership roles. I support organisations to achieve gender diverse leadership. I advocate for equitable gender diversity in STEM leadership to government and private STEM organisations. I have slowly started to disrupt and change the narrative and experience around STEM leadership.

What I have learnt along the way is that my journey over the past twenty-five years, and the challenges I have faced, have brought me to the vision I have today. That journey has provided the skills and experiences to achieve my vision in a unique way. I am now fulfilling the desire to create long-lasting impact and change that manifested subconsciously in my early twenties in a way I never imagined. I want to leave this world a better place and am doing that by empowering women in STEM to lead.

Dr Susan McGinty

Dr Susan McGinty is the Founder and Director of Aya Leadership. She is an award-winning scientist and highly regarded leader and leadership coach, with 25 years' experience in STEM, Defence and National Security, who is inspired to transform the profile of leadership in STEM and security-related organisations.

Inspired by the challenges of her own leadership journey, Susan is passionate about increasing gender diversity in STEM leadership. Susan founded Aya Leadership as a specialised leadership development partner for STEM and security organisations. She works with organisations to identify and deliver the development strategies that will build leadership gender equity; and works with women to develop enduring and impactful leadership via tailored leadership development programs.

Susan has a PhD in Medicinal & Organic Chemistry and a Master of Leadership. Following ten years in research, Susan spent 13 years in corporate leadership positions in Defence and National Security leading a range of specialised scientific and technical teams, managing key international Defence and national security partnerships, and advising on key domestic and international scientific policies.

Susan is a certified coach with fifteen years' experience in adult education, coaching and mentoring in the STEM, security, university, research, corporate, government and community sectors. She was identified by The Australian Business Journal as one of the 20 Australian leadership experts to watch in 2021.

Find out more about how Aya Leadership is transforming the profile of leadership in STEM and uplifting and empowering more STEM women into leadership roles, and how its range of leadership development programs for emerging and established female STEM leaders accelerates enduring and impactful leadership growth at https://ayaleadership.com.

CHAPTER 18

Shatter the Glass – The Future of STEM is Feminine!

Tiffiny J. Roper

Having been in the IT world about twenty-five years overall, originally starting in Sales (just to get my foot in the door as a new graduate of college at the time), then moving to Proposals and Project Management, I often struggled to find my footing as a woman in the very male-dominated IT world. I found there were specific boxes women were usually put in: the administrative role or the sales role, where it was okay to act more feminine. Then, there were all the rest of the roles typically held by men, where the women were expected to act more masculine to fit in or to have any hopes of promotions.

When I was in Sales, I was either the top Sales Representative or at the very top with one or two men. In almost five years, I never missed quota and was never below the top three in Sales for the entire floor, generally number one. However, I saw how frequently the men in the top with me would become threatened by my success. When this occurred, I also noticed how I was treated by these men. Instead of it being a friendly competition between co-workers, it quickly became more like a war zone. Some of these men would find ways to try to steal my orders, though it was common that if one person worked up a quote for a customer, that it would go back to that Sales Representative, unless they were out of the office. Then you would put the order in their name so they would get credit.

For me, though, that generally did not happen. These particular men felt they had to beat me or else. Instead, they would take credit for the orders, even when I was sitting next to them, available to take the order myself. Other men would get some of the female sales representatives to start rumors, such as saying I was dating someone in the workplace, even though I was engaged then married during this time. It became obvious my future wasn't going to be in Sales. Getting the best territories or being promoted to Sales Manager stemmed from the relationships built on the golf course, and my outstanding sales metrics didn't seem to matter.

So, after five years, I decided to move out of Sales, yet found I was still in the good old boys' club. I felt like I was pledging a fraternity as a woman, and nobody was interested. I quickly figured out the first part of playing the corporate game. They would set up interviews, because they were required to do a certain number of them, but they had already had people on the bench they were planning to hire. Those people always were male. After several internal interviews for different positions, which could go on for hours and be very exhausting, I saw others hired instead of me. It became obvious I was on the wrong side of those being hired.

After working at this company a few years, I'd often heard them talk about thinking outside the box, but I had seen that they really only wanted these "yes men" in their management line. In fact, there were so many women that were refused manager and director roles, and I knew of several, personally, who decided to leave the company and ended up with much higher positions elsewhere. It was so bad, the company ended up being sued by a couple of women, and they found enough discrimination against these women that all the women under their employ, at the time, won a small settlement, along with an NDA regarding the settlement details.

After many interviews, I eventually met with a woman who was hiring for Proposal Managers, people that reply to customers bids wanting to buy from your company or another, and you had to respond to their questions, then they would decide what company to use for their purchase. I believe, because it was a woman hiring (one of the few women in manager, director, or VP roles at a Fortune 100 company with over 100,000 employees), that's why I had a chance at this position, and after several interviews, I was hired. I was so excited!

Sadly, the excitement didn't last long. Soon, I watched people judge my female manager, whom I really respected as she was a great leader. They called her emotional because she was open with how she felt about things. It was their way, the fraternity, boys' club way, to dismiss women who didn't play by the "yes men" rules to fit in, and instead remove this threat, the powerful woman in charge.

This taught me if I wanted to be taken seriously in this IT world, I needed to take on more masculine type behaviors, such as: never expressing feelings or showing emotions, pushing things with force, being more critical, and being the typical "yes men" that were prevalent in leadership positions at this company. Instead of standing out, and rising above others, by being the beautiful emotional feminine beings these women and I were designed to naturally be, I saw myself, and my female colleagues, shut that side of their personalities down, resulting in not being their authentic, true selves. I felt like I was wearing a mask, one that didn't quite fit right. Soon, I couldn't recognize myself anymore.

Finally, I started learning the importance of who you know, since building relationships in this environment, and anywhere, is key to getting where you want to go in your career. So, I quickly started to reach out to people I wanted to network with and decided, why not start at the top? I then messaged some top Directors and VPs of the company to start discussing my career overall. Most of them did take the time out of their busy schedules to meet with me, and it was really appreciated. During these meetings, I always had a list of questions to ask them about my career and where they could see someone with my experience, education, and interests in the company's future. I would also ask their advice on what steps they recommended I take, or what I should learn, and I would make sure to follow through with these actions. I'd follow-up with them directly, so they knew I took our meetings seriously.

Lastly, I would ask for the names of three other people, Directors or Managers, that may not have taken the time to meet with me without their referral. They recommended I meet with people based on their thoughts of my future at the company. I was able to meet with so many great people and kept the networking going for the next several years. I would also stay in touch with these people to continue to build the relationships I started with those meetings. That's the power of networking and not being afraid to hear the word no. Sometimes when we get no's, we are just asking the wrong people, so just keep asking.

Once I had built some of these relationships, I started to ask these people, those that I strongly connected with, if they could mentor me. That's how I found the power of having mentors, and not just one, but several. I started helping others with the information I learned and really enjoyed it. In over ten years of being there, it was truly the first time I started eventually feeling empowered in this environment, at that point.

Later, I decided to move out of Proposals and ended up in Project Management (a career I first heard about from one of the VPs I'd met with) and did it for twenty years. I was finally in a position where I was respected and listened to (because I met with these mentors regularly and followed their advice), and I finally felt comfortable being myself. Being truly me: the intuitive, creative, loving, nurturing, giving feminine force that enjoyed going with the flow and being different than the men around me.

I eventually left that company, after almost fifteen years, and continued my career in Project Management elsewhere. I went through a lot at this company, and eventually realized it wasn't the company I wanted to support in the long-term. Too many women were still not in the positions they deserved when I left. I do honor that company for teaching me so many lessons, lessons that I still use today.

Out of all the lessons I learned through the years, the biggest thing I've learned is to not be afraid to really BE ME! You are the ONLY YOU in the entire universe, and you are exactly perfect how you are right now. We all have things we want to work on, it's why we are here. Just know you are already amazing and exactly where you need to be. So, instead of trying to fit into a mold, feel empowered to be exactly who you are. Then, all you have to do is sell you, as your authentic self, in your next interview. The right company, and the right hiring managers, want someone that is real, that they can connect to, that will connect to their team and customers.

It's also about how you treat others in the world, and that starts with how you treat yourself. Then, be your true self and start networking and building relationships and finding mentors. Don't fear asking for advice or help from others, because you will get a yes more than you'd think. People are, overall, good and do want to help.

Truly, women are at their best when they are their authentic feminine selves. If that's their energy, as some women do identify with more of a masculine energy, and that's fine for them because it's authentic to them. Just don't let anyone make you feel wrong for being authentically feminine, authentically you. Instead, shatter that glass ceiling, then reach back and help others around you, so they can shatter that glass ceiling too. We are not each other's competitors as women. That's a lie we have been told for years. We should hold each other up, instead, and truly want the success of those women around us, just as we want our own success. There is room for us all.

Remember, in STEM, women need to do the exact opposite of what we have been doing for years, if we want to truly succeed. We need to focus on our amazing feminine attributes and the strength we get from being feminine, instead of trying to fit into the good old boys' club. Women should utilize their creative minds, intuitive gut feelings, and their emotions to build critical relationships and network. We also need to make the tough, but important, business decisions and come up with customized solutions that are needed in order to drive to the top of the corporate IT world and break that glass ceiling once and for all.

Now, I challenge you to be the inspirational woman that succeeds, by using these amazing feminine attributes to have the ideal career you've desired and by your design, a career you can be passionate about and feel like you are living a life of purpose, while truly embodying your whole, feminine self. Go support and mentor other women in their growth, and teach them the steps you learned to move up the corporate ladder. The STEM world needs this feminine energy, in all its glory, to drive the business world to the next level. Now is the time, and the world is ready. The future in STEM is FEMININE! See you at the top!

Tiffiny Roper

Tiffiny Roper is on a mission to create amazing female leaders of tomorrow that this world desperately needs. She accomplishes this as a Life Coach for Moms of young daughters, working passionately with them to reach their goals and becoming the best role models they can be. In doing so, they live with a new purpose-driven life filled with joy and inspire those around them to do the same, starting with their daughters. She uses her twenty years of Project Management experience to keep Moms accountable in hitting their goals. She's also a speaker and best-selling author that loves creating memories with her husband of twenty years and young twin daughters, including coaching their softball team and leading their Girl Scout troop.

Connect with Tiffiny at tiffiny@girlmomcoaching.com.

Endnotes

1 https://www.isc2.org/Research/Women-in-Cybersecurity#

2 https://www.mckinsey.com/featured-insights/diversity-and-inclusion/seven-charts-that-show-covid-19s-impact-on-womens-employment

3 https://www.isc2.org/-/media/ISC2/Research/2021/ISC2-Cybersecurity-Workforce-Study-2021.ashx

Manufactured by Amazon.ca
Bolton, ON